LC Circuits

by

Rufus P. Turner, Ph. D.

Howard W. Sams & Co., Inc.
4300 WEST 62ND ST. INDIANAPOLIS, INDIANA 46268 USA

International Standard Book Number: 0-672-21694-9
Library of Congress Catalog Card Number: 79-57616

Printed in the United States of America.

Preface

Technologically and historically, the familiar combination of inductance and capacitance, the LC circuit, is the basic selective unit of electronics. Originally delegated to the tuning of radio apparatus, the LC circuit has found application far afield in many areas of electronics.

This book describes a number of practical LC circuits and offers enough background theory to promote the understanding of them. A sufficient amount of space has been devoted also to resistance, since that property is inherent in practical inductors and capacitors.

Although the material is addressed to the electronics student, technician, and experimenter, more advanced readers may find certain parts of it useful, if only for reference purposes. A minimum of mathematics is employed—physical explanations being preferred where feasible—and frequent illustrative examples demonstrate the necessary calculations.

I hope that this book will serve both the novice and the virtuoso.

RUFUS P. TURNER

Contents

CHAPTER 1

CHAPTER 2

CHAPTER 3

CHAPTER 4

APPENDIX A

APPENDIX B

APPENDIX C

Fundamental Theory

This chapter digests those parts of basic electronics that are essential to an understanding of inductance-capacitance (LC) circuits. These are specific items requiring for their understanding a general familiarity with electrical theory and the mathematics of electronics, and it is assumed that the reader has that background.

1.1 THE AC CYCLE—RATE OF CHANGE

Fig. 1-1A depicts a sinusoidal ac voltage in terms of a vector rotating at constant velocity. The magnitude of this vector is E_{max}, the maximum value attained by the ac voltage. As the vector moves in a counterclockwise direction from its starting point at the horizontal axis, it generates an angle θ which increases from initial zero to 360° (2π radians) in each complete rotation. If the figure is drawn to scale, the instantaneous voltage, e, is depicted by the length of the half chord extending from the tip of the vector to the horizontal axis. (Although the vector is rotating at constant angular velocity, the length of this half chord does not change at a constant rate; see Table 1-1.)

From Fig. 1-1A, it is easily seen that the instantaneous voltage is zero at 0°, since here the half chord has no length at all, and is maximum at 90°, since here the half chord has its maximum length. Thus, instantaneous voltage e starts at zero, increases to the maximum positive value ($+E_{max}$) at 90° ($\pi/2$ radians), returns to zero at 180° (π radians), increases to

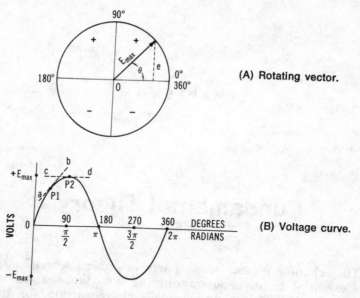

(A) Rotating vector.

(B) Voltage curve.

Fig. 1-1. Development of sine wave.

the maximum negative value $(-E_{max})$ at 270° ($3\pi/2$ radians), and returns to zero at 360° (2π radians). At this latter point, the vector has traced out a complete cycle, and a new cycle begins with the continued rotation of the vector. The length of line e, and accordingly the value of the instantaneous voltage, is proportional to the sine of angle θ, and for a circle of

Table 1-1. Voltage Change and Rate Change

Rotation (degrees)	Voltage Change* (V)	Rate of Voltage Change (V/degree)
0–10	0–0.174	0.0174
10–20	0.174–0.342	0.0168
20–30	0.0342–0.5	0.0158
30–40	0.5–0.643	0.0143
40–50	0.643–0.766	0.0123
50–60	0.766–0.866	0.01
60–70	0.866–0.940	0.0074
70–80	0.940–0.985	0.0045
80–90	0.985–1	0.0015

* $E_{max} = 1$ V

unit radius (i.e., $E_{max} = 1$), it is *equal* to sin θ. The value of instantaneous voltage at any point, therefore is:

$$e = E_{max}\sin \theta \qquad (1\text{-}1)$$

Fig. 1-1B shows the familiar plot of instantaneous voltage versus angle. This curve is identical with that of the sine function from trigonometry, hence the term *sine wave* (or *sinusoidal*).

Illustrative Example: The 115-V power-line voltage has a maximum (90°) value of 162.6 volts. Calculate the instantaneous value at 60°. Sin 60° = 0.866. From Equation 1-1, $e = 162.6(0.866) = 140.8$ V.

A close examination of Fig. 1-1 shows that not only is the instantaneous voltage continuously changing during the cycle, but the *rate* of change also is changing. (See Column 3 in Table 1-1.) This can be shown graphically by drawing tangents to the curve at points of interest and observing the slope of the tangent. Thus, in Fig. 1-1B, the tilt of tangent *ab* at Point P1 (45°) reveals a moderate rate of change, whereas tangent *cd* at Point P2 (90°) has no tilt whatever and shows that there is no change at all at this point. In the sine-wave cycle, the *rate* of change thus is zero at 90° and 270° when the cycle is passing through its maximum points, and is maximum when the cycle is passing through its zero points (0°, 180°, and 360°).

This can be shown mathematically: (1) The increment of voltage around the point of interest divided by the increment of the angular rotation at that point equals the slope; thus, slope $= \Delta E/\Delta\theta$. From Fig. 1-1B, it is easily seen that at 90° (Point P$_2$), $\Delta E = 0$, and the slope and rate of change accordingly are zero, whereas at 180° ΔE is very large for a small value of $\Delta\theta$, so the slope is steep at this point and the change is great. (2) From calculus, the rate of change of a sine wave at a point of interest is proportional to the cosine of the angle at that point. The cosine of 90° or of 270° is zero, so the rate of change is zero at those points. The cosine of 0° and of 360° is 1, so the rate of change is maximum at those points where the cycle crosses the zero line (the maximum value of the cosine function is 1).

At a given frequency f, the voltage vector describes $2\pi f$ radians per second (where f is in hertz, and $\pi = 3.1416$). The expression $2\pi f$ is often represented by lowercase Greek omega: ω. (Appendix A gives values of ω at a number of frequencies between 1 Hz and 100 MHz.) The maximum rate of change in voltage is equal to $2\pi fE_{max}$ (also written ωE_{max}), and it follows from the previous discussion that this rate of change must occur at zero-voltage points in the cycle. At every point, the rate of change may be found by multiplying the maximum rate of change by the cosine of the angle at the point of interest:

$$\text{Rate of Change} = 2\pi fE_{max}\cos\theta = \omega E_{max}\cos\theta \quad (1\text{-}2)$$

For example, in a 500-Hz cycle having a maximum voltage of 15 V, the rate of change at 65° (where $\cos\theta = 0.42262$) is

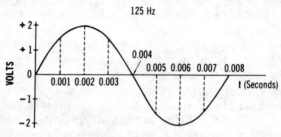

(A) Typical cycle.

t	θ			
(seconds)	(degrees)	(radians)	sin θ	cos θ
0	0	0	0	1
0.001	45	0.7854	0.7071	0.7071
0.002	90	1.5708	1	0
0.003	135	2.3562	0.7071	−0.7071
0.004	180	3.1416	0	−1
0.005	225	3.9270	−0.7071	−0.7071
0.006	270	4.7124	−1	0
0.007	315	5.4978	−0.7071	0.7071
0.008	360	6.2832	0	1

(B) Numerical relationship.

Fig. 1-2. Illustrative cycle.

$2(3.1416)500(15)0.42262 = 19,915$ volts per degree. At 0° (where $\cos\theta = 1$), the rate of change is $2(3.1416)500(15)1 = 47,124$ volts per degree. At 90° (where $\cos\theta = 0$), the rate of change is $2(3.1416)500(15)0 = 47,124 \times 0 = 0$ volts per degree.

Fig. 1-2A illlstrates a specific example. This is one cycle of 125-Hz ac voltage having a maximum value of 2 volts. The horizontal axis is divided into seconds of time, instead of degrees, but these time units are easily converted into an angle, if desired:

$$\theta = 2\pi ft = \omega t \tag{1-3}$$

where,

θ is the angle in radians,
f is the frequency in hertz,
t is the time in seconds,
π is 3.1416,
ω is $2\pi f$.

To convert θ to degrees, multiply by 57.296. To obtain degrees directly, change Equation 1-3 to:

$$\theta = 360 \, ft \tag{1-4}$$

Fig. 1-2B shows the relationship between selected angles and time instants, and gives sines and cosines of these angles. Note the rate of change exhibited in this cycle. The maximum rate of change here ($2\pi fE_{max}$) is 1570.8 V/s. From Equation 1-2, the rate of change at 0.004 second (180°, π radians) is $2(3.1416)125(2)(-1) = 1570.8(-1) = -1570.8$ V/s; the rate of change at 0.001 s (45°, $\pi/4$ radians) is $2(3.1416)125(2)$ $0.7071 = 1110.7$ V/s; and the rate of change at 0.002 s (90°, $\pi/2$ radians) is $2(3.1416)125(2)0 = 1570.8 \times 0 = 0$.

It should be clear by now that the rate of change in voltage at a selected point in the ac cycle is markedly different in magnitude from the instantaneous voltage at that point. In the preceding example, for instance, it is seen that the instantaneous voltage at the 0.001-second point (45°) is (from Equation 1-1) $2 \times \sin 45° = 2(0.7071) = 1.414$ V; but the rate of change of voltage at that point (from the foregoing paragraph) is 1110.7 volts per second. The various aspects of rate of change in the ac cycle must be mastered and remembered, since a clear understanding of inductor and capacitor operation and of LC circuits demands this comprehension.

What has been said in this section about ac voltage and the voltage cycle applies equally well to ac current and the current cycle. It is necessary only to substitute the word *current* for voltage, and the symbols I and i for E and e in the discussion.

1.2 NATURE OF RESISTANCE

It is necessary to introduce a discussion of resistance at this point, since internal resistance is unavoidable in inductors and capacitors and can influence the performance of LC circuits.

Resistance (R or r) is the simple opposition offered to the flow of an electric current. It is somewhat analogous to the friction encountered by flowing water. Resistance is directly proportional to voltage and is inversely proportional to current, as shown by Ohm's law:

$$R = \frac{E}{I} \tag{1-5}$$

where,
  R is the resistance in ohms,
  E is the voltage in volts,
  I is the current in amperes.

An applied voltage E thus will force a current $I = E/R$ through a resistance R; likewise, a current I through a resistance R will produce a voltage drop $E = IR$ across the resistance. While all conductors of electricity exhibit some resistance, however tiny, resistance is primarily the property of *resistors*. Resistance is measured in ohms and in multiples and submultiples of the ohm (See Table 1-2).

Conventional (ohmic) resistors are made principally from suitable metals in the form of wire, strip, ribbon, or film; from carbon or graphite; and from certain controllable-resistivity compositions, mixtures, and oxides. Nonlinear (non-ohmic) resistors, whose resistance depends upon applied voltage, and thermistors, whose resistance depends upon temperature, are made from semiconductors or from suitable compounds, such as complex oxides. In addition to discrete resistors, there are integrated resistors which are processed into integrated circuits (ICs). Variable, as well as fixed, resistors are readily obtainable, and all resistors are available in a wide range of resistance and power ratings, shapes, and sizes. For special purposes, the internal resistance of some other device

Table 1-2. Units of Resistance

Unit	Number of Ohms	Symbol
Microhm	1×10^{-6}	$\mu\Omega$
Milliohm	1×10^{-3}	$m\Omega$
Ohm	1	Ω
Kilohm	1×10^{3}	$k\Omega$
Megohm	1×10^{6}	$M\Omega$
Gigohm	1×10^{9}	$G\Omega$
Teraohm	1×10^{12}	$T\Omega$

—such as a vacuum tube, transistor, semiconductor diode, or filament-type lamp—is sometimes employed instead of a resistor *per se.*

Pure resistance introduces no phase shift. Fig. 1-3 illustrates this zero-phase-shift feature; in both the wave plot (Fig. 1-3B) and the vector diagram (Fig. 1-3C), current and voltage are in step at all points. Because pure resistance is impossible to attain in practice, all resistors possess some inherent inductance and capacitance, but these extraneous properties are usually, but not always, so minute that they can be ignored.

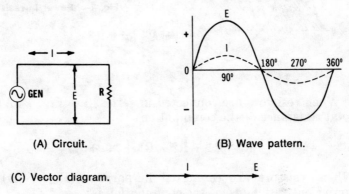

(A) Circuit.

(B) Wave pattern.

(C) Vector diagram.

Fig. 1-3. Phase relationship: resistance.

Resistance is a dissipative property: that is, resistors consume power. Current (I) flowing through resistance (R) causes heat to be generated in the resistor, just as mechanical friction gives rise to heat in bodies rubbed together. In this way, electrical energy is converted into heat energy, and the latter represents energy that is lost for some intended electrical work. The electrical loss is *power dissipation* and may be expressed as:

$$P = I^2R = \frac{E^2}{R} \qquad (1\text{-}6)$$

where,
    P is the power loss in watts,
    I is the current through the resistance in amperes,
    E is the voltage across the resistance in volts,
    R is the resistance in ohms.

The sine-wave pattern shown in Fig. 1-4 shows resistor power in relation to current and voltage.

When operated within their ratings, good-grade conventional resistors undergo very little change in resistance as a result of variations in voltage, temperature, or frequency. Exceptions are voltage-dependent resistors (vdr's), which are designed to be voltage sensitive, and thermistors, which are designed to be temperature sensitive.

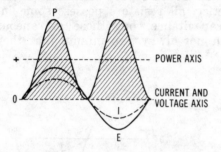

POWER AXIS

CURRENT AND
VOLTAGE AXIS

Fig. 1-4. Power in resistor.

When resistors are connected in series (see Fig. 1-5A), the total resistance of the combination is:

$$R_t = R1 + R2 + R3 + \ldots + R_n \qquad (1\text{-}7)$$

When resistors are connected in parallel (Fig. 1-5B), the equivalent resistance of the combination is:

$$R_{eq} = \cfrac{1}{\cfrac{1}{R1} + \cfrac{1}{R2} + \cfrac{1}{R3} + \ldots \cfrac{1}{R_n}} \qquad (1\text{-}8)$$

If only two resistors are connected in parallel, the equation can be simplified to $R_{eq} = (R1R2)/(R1 + R2)$. If n resistors,

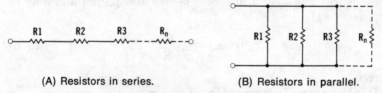

(A) Resistors in series. (B) Resistors in parallel.

Fig. 1-5. Basic resistor circuits.

each having the same value R, are connected in series, the total resistance of the combination is $R_t = nR$. If n resistors, each having the same value R are connected in parallel, the equivalent resistance of the combination is $R_{eq} = R/n$.

Table 1-3. Units of Inductance

Unit	Number of Henrys	Symbol
Henry	1	H
Millihenry	0.001	mH
Microhenry	1×10^{-6}	μH
Nanohenry	1×10^{-9}	nH
Picohenry	1×10^{-12}	pH

1.3 NATURE OF INDUCTANCE

Inductance (L) is the property exhibited by a conductor or by a coil of wire (inductor) which retards the buildup of current when a voltage is applied, or the decay of current when a voltage is removed. It is sometimes called *electrical inertia*, and is caused by the counter emf ("back voltage") produced by the magnetic field surrounding the inductor.

Inductance is measured in henrys (H); but since this is a large unit, submultiples of the henry are often used (see Table 1-3). While inductance is present in any conductor, even in a straight wire, it is primarily a property of inductors (also called *coils*). A simple practical inductor consists of a coil whose turns are wound in a single layer on a nonmetallic, cylindrical form and its inductance (L) depends upon the number of turns (N), the length of the coil (l), and the diameter of the coil (d):

$$L = \frac{0.2 \, d^2 \, N^2}{3d + 9l} \tag{1-9}$$

where,
    L is the inductance in microhenrys,
    d is the diameter of the winding in inches,
    l is the length of the winding in inches,
    N is the number of turns.

The inductance formula becomes somewhat different when the inductor is wound in several layers or when it is wound on a core of magnetic material (iron, steel, ferrite, etc.). The core and the multilayer winding (increased number of turns) each increases the inductance (the higher the permeability—μ—of the core material, the fewer the turns needed for a given inductance). Inductors are manufactured in a wide range of inductance, current, voltage, and internal-resistance ratings and in numerous shapes and sizes. Inductors are available in fixed and variable types.

15

Inductance is also found in places other than in inductors. It is inherent in other components and devices than inductors. For example, by their nature transformer windings have inductance, and so do the leads of capacitors, resistors, and other components. Even short, straight wires exhibit inductance, which—though tiny—can be deleterious at very high frequencies.

All inductors have internal resistance (R) which at dc and low frequencies is due to the resistance of the wire. At higher frequencies, other factors—such as core losses and skin effect —combine with the wire resistance to determine the total resistance of an inductor. In a high-quality inductor, resistance, which acts in series with the inductance, is held to a minimum.

When a dc voltage is applied to an inductor, the resulting final current is determined by the resistance of the winding. But the current does not reach this value at once, nor can it be rapidly increased or decreased. Any such change is opposed by the counter emf, which is opposite in polarity and is produced by the surrounding magnetic field. This means that the current does not even begin to flow until a short time after application of the voltage. Thus, current lags voltage (voltage leads current) in an inductive circuit.

Because of the action just described, pure inductance would introduce a 90° phase shift in an ac circuit. Fig. 1-6 illustrates this 90°-lagging phase feature. In both the wave plot (Fig. 1-6B) and the vector diagram (Fig. 1-6C), current and voltage are out of step by 90° at all points. Examination of Fig. 1-6B shows that maximum inductor current flows when the rate of change in voltage is maximum (that is, when the ac cycle is passing through zero) ; and, conversely, inductor current is zero when the rate of change in voltage is zero (that is. when the ac cycle is at maximum). For a clarification, see Section 1.1 for a discussion of rate of change. Because pure inductance is impossible to attain in practice, all inductors possess some inherent resistance and capacitance and these extraneous properties are usually small. Nevertheless, inherent resistance (internal losses) prevents a practical inductor from introducing *full* 90° phase shift.

Pure inductance, unlike resistance, consumes no power. This results from energy being stored in the magnetic field during one half-cycle (when the field is expanding) and being returned to the ac generator during the next half-cycle (when the field is collapsing). The wave pattern in Fig. 1-7 depicts inductor power (for an ideal inductor) in relation to current and voltage (compare with Fig. 1-4, the wave pattern for

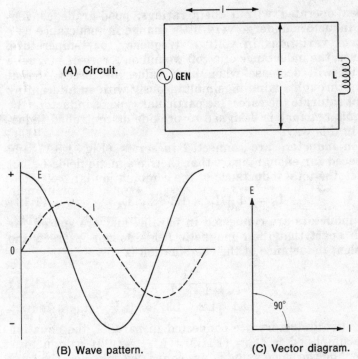

(A) Circuit.

(B) Wave pattern.

(C) Vector diagram.

Fig. 1-6. Phase relationship: inductance.

resistor power, and Fig. 1-11, the wave pattern for capacitor power). Note that the frequency of the power wave is double that of the current wave or the voltage wave. In a practical inductor, the only power loss is that associated with the internal resistance of the inductor.

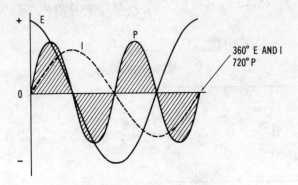

360° E AND I
720° P

Fig. 1-7. Power in ideal inductor.

When operated within their ratings, good-grade conventional inductors undergo very little change in inductance as a result of variations in voltage, frequency, or temperature. However, the inductance of a coil wound on a core of magnetic material will decrease when high values of direct current flowing through a winding simultaneously with an alternating current saturate the core. One particular type of inductor (the *saturable reactor*) is designed to provide dc-controlled inductance in this way.

When inductors are connected in series (Fig. 1-8A) and are spaced far enough apart that their magnetic fields do not interact, the total inductance of the combination is:

$$L_t = L1 + L2 + L3 + \dots L_n \qquad (1\text{-}10)$$

When inductors are connected in parallel and are spaced far enough apart that their magnetic fields do not interact, the equivalent inductance of the combination is:

$$L_{eq} = \cfrac{1}{\cfrac{1}{L1} + \cfrac{1}{L2} + \cfrac{1}{L3} + \cdot\cdot \cfrac{1}{L_n}} \qquad (1\text{-}11)$$

If only two inductors are connected in parallel, the equation can be simplified to $L_{eq} = (L1L2)/(L1 + L2)$. If n inductors, each having the same value L, are connected in series, the total inductance of the combination is $L_t = nL$. If n inductors, each having the same value L, are connected in parallel, the equivalent inductance of the combination is $L_{eq} = L/n$.

The property which has been under discussion up to this point is strictly termed *self-inductance*. When two inductors are so close physically that their fields overlap (that is, the coils are coupled), a second variety of inductance—*mutual inductance* (M)—comes into play. Interaction occurs because

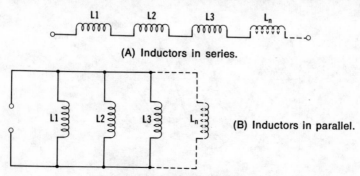

(A) Inductors in series.

(B) Inductors in parallel.

Fig. 1-8. Basic inductor circuits.

the first coil induces a counter emf into the second coil and vice versa (the effect differs from that resulting from the connection of two *well-spaced* inductors in series). When the two inductors are so tightly coupled (coefficient of coupling $k = 1$) that all the lines of force from one inductor link all the turns of the other, M is maximum, and when the two inductors are completely decoupled (that is, no flux from one interacts with the other), M is zero.

1.4 NATURE OF INDUCTIVE REACTANCE

Fig. 1-9 shows the relationship between sinusoidal voltage and current for a pure inductance. Here, E_i is the applied voltage, I is the resulting current, and E_c is the counter emf. This counter voltage opposes the applied voltage, since—as is seen in Fig. 1-9—the two are out of phase with each other.

The counter voltage increases and the inductor current accordingly decreases as the frequency of the applied voltage increases, and vice versa. At a given frequency, the counter voltage increases and the inductor current therefore decreases as the inductance increases, and vice versa. An inductor therefore offers frequency-dependent opposition to the flow of alternating current. By means of calculus, it can be shown that this opposition is equal to ωL. This opposition is termed *inductive reactance*, is measured in ohms, and is given by:

$$X_L = \omega L = 2\pi fL \qquad (1\text{-}12)$$

where,

X_L is the inductive reactance in ohms,
L is the inductance in henrys,
f is the frequency in hertz,
ω is $2\pi f$,
π is 3.1416.

The relations between current, voltage, and reactance are expressed in a form often called *Ohm's law for ac*:

$$X_L = \frac{E}{I}, \quad E = IX_L, \quad \text{and } I = \frac{E}{X_L} \qquad (1\text{-}13)$$

where,

X_L is in ohms,
I is in amperes,
E is in volts.

From Equation 1-13, it is seen that an alternating current flowing in an inductive reactance produces a voltage drop

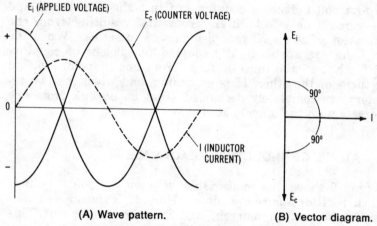

(A) Wave pattern.   (B) Vector diagram.

Fig. 1-9. Inductor current/voltage relationship.

IX_L; and, because of the phase of inductive reactance, this voltage leads the current by 90°.

Inductive reactance is a major factor in all LC circuits. It, like capacitive reactance (Section 1.6), is the property that imparts frequency sensitivity to these circuits. For a given inductance, doubling the frequency doubles the reactance, halving the frequency halves the reactance, and so on. Similarly, for a given frequency, halving the inductance, halves the reactance, doubling the inductance doubles the reactance, and so on. Appendix B gives the 1000-Hz reactance corresponding to ten inductances spaced in decade relationship from 1 μH to 1000 H.

1.5 NATURE OF CAPACITANCE

Capacitance (C or c) is the ability to store an electric charge. It is somewhat analogous to the ability of a tank to store a quantity of fluid. Capacitance is measured in farads (F); but since this is an extremely large unit, submultiples of the farad generally are used (see Table 1-4). While capacitance exists between any two nearby conductors, it is primarily a property of *capacitors*. A simple practical capacitor consists of two identical metal plates or films separated by a dielectric (air, solid insulant, or liquid insulant), and its capacitance (C) is directly proportional to the area (A) of one plate and to the dielectric constant (k) of the dielectric, and is inversely proportional to the thickness (t) of the dielectric:

$$C = \frac{kA}{4.45\,t} \qquad \text{(1-14)}$$

where,
C is the capacitance in picofarads,
k is the dielectric constant,
A is the area of one plate in square inches,
t is the thickness of the dielectric in inches.

Capacitors may be of the two-plate type or multiple-plate type. Both are manufactured in fixed and variable varieties and are offered in a wide range of capacitance and voltage ratings, shapes, and sizes. Numerous dielectrics (air, paper, oil, ceramic, mica, glass, plastic, etc.) are employed; the higher the dielectric constant, the smaller the required area and thickness for a given capacitance. In addition to discrete capacitors, there are integrated capacitors which are processed into integrated circuits (ICs). For special purposes, the internal capacitance of some other device—such as a vacuum tube, semiconductor diode or rectifier, or transistor—sometimes is employed instead of a capacitor *per se*.

Table 1-4. Units of Capacitance

Unit	Number of Farads	Symbol
Farad	1	F
Microfarad	1×10^{-6}	μF
Nanofarad	1×10^{-9}	nF
Picofarad	1×10^{-12}	pF

When a dc voltage is applied to a capacitor, the latter becomes charged because energy then is stored in the electric field between the plates. When the voltage is disconnected, the capacitor retains the charge, the original voltage E appearing across the capacitor. If the capacitor were perfect, it would remain charged indefinitely. When an external circuit is connected across the charged capacitor, the capacitor discharges through the circuit and its voltage falls to zero. Thus, a current flows into the capacitor to charge it, and out of the capacitor in the opposite direction to discharge it. No current can flow *through* the capacitor, because of the dielectric between the plates.

When a voltage (E) is applied to a given capacitor, the quantity of charge (Q) that the capacitor receives is directly proportional to the voltage and to the capacitance (C):

$$Q = CE \qquad \text{(1-15)}$$

where,

Q is the quantity of charge in coulombs,
C is the capacitance in farads,
E is the voltage in volts.

When the voltage first is applied, a heavy current flows into the capacitor and the voltage across the capacitor is low. This current diminishes with time while the capacitor voltage increases. When the capacitor becomes fully charged, the current ceases and the capacitor voltage then is equal to the charging voltage. Thus, the capacitor current leads the capacitor voltage. When an ac voltage is applied to a capacitor, charging current flows into the capacitor during one half-cycle, and discharging current flows out of the capacitor during the other half-cycle; so an alternating current flows in the circuit, but—again—not *through* the capacitor.

Because current leads voltage in a capacitor, the two are out of phase in a capacitor circuit. Pure capacitance would introduce a 90° phase shift in an ac circuit. Fig. 1-10 illustrates this 90°-leading phase feature. In both the wave plot (Fig. 1-10B) and the vector diagram (Fig. 1-10C), current and voltage are 90° out of step, with current leading at all points.

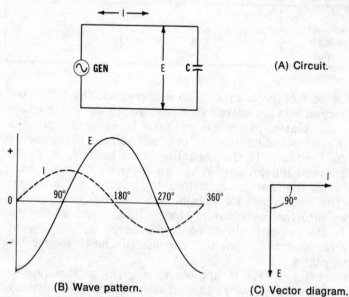

(A) Circuit.

(B) Wave pattern.

(C) Vector diagram.

Fig. 1-10. Phase relationship: capacitance.

Examination of Fig. 1-10B shows that maximum capacitor current flows when the rate of change in voltage is maximum (i.e., when the ac voltage cycle is at zero); and, conversely, capacitor current is zero when the rate of change in voltage is zero (i.e., when the ac voltage cycle is at maximum). Here, $I = C(de/dt)$. In this connection, see Section 1.1 for a discussion of rate of change. Because pure capacitance is impossible to attain in practice, all capacitors possess some inherent resistance and inductance, but these extraneous properties are usually tiny. Nevertheless, inherent resistance (internal losses) prevents a practical capacitor from introducing *full* 90° phase shift.

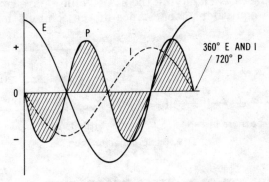

Fig. 1-11. Power in ideal capacitor.

Pure capacitance, unlike resistance (but like pure inductance), consumes no power. This is because energy stored in the electric field during charge is returned to the ac generator during discharge. The sine-wave pattern in Fig. 1-11 depicts capacitor power in relation to current and voltage (compare with Fig. 1-4, the wave pattern for resistor power, and Fig. 1-7, the wave pattern for inductor power). In a practical capacitor, the only power loss is that associated with the internal resistance of the capacitor, and this is small in a high-grade capacitor.

When operating within their ratings, good-grade conventional capacitors undergo very little change in capacitance as a result of variations in voltage, temperature, or frequency. Exceptions are voltage-variable capacitors (varactors), which are designed to be voltage sensitive, and compensating capacitors, which are designed to be temperature sensitive.

When capacitors are connected in series (Fig. 1-12A), the equivalent capacitance of the combination is:

$$C_{eq} = \frac{1}{\frac{1}{C1} + \frac{1}{C2} + \frac{1}{C3} + \ldots + \frac{1}{C_n}} \qquad (1\text{-}16)$$

When capacitors are connected in parallel (Fig. 1-12B), the total capacitance of the combination is:

$$C_t = C1 + C2 + C3 + \ldots + C_n \qquad (1\text{-}17)$$

If only two capacitors are connected in series, the equivalent capacitance of the combination is $C_{eq} = (C1C2)/(C1 + C2)$. If n capacitors, each having the same value C, are connected in parallel, the total capacitance of the combination $C_t = nC$. If n capacitors, each having the same value C, are connected in series, the equivalent capacitance of the combination is $C_{eq} = C/n$.*

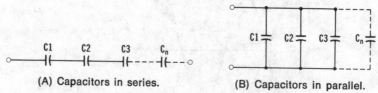

(A) Capacitors in series.
(B) Capacitors in parallel.

Fig. 1-12. Basic capacitor circuits.

1.6 NATURE OF CAPACITIVE REACTANCE

Fig. 1-13 shows the relationship between the sinusoidal voltage and current for a pure capacitance. Here, E_i is the applied voltage, I is the resulting current, and E_c the counter voltage. The latter is the voltage drop across the capacitor and is analogous to the counter emf developed by an inductor carrying alternating current. The counter voltage, E_c, opposes the applied voltage, since—as is seen in Fig. 1-13—the two are out of phase with each other.

The counter voltage decreases and the capacitor current accordingly increases as the frequency of the applied voltage increases, and vice versa. For a given frequency, the counter voltage decreases and the capacitor current therefore increases as the capacitance increases, and vice versa. A capacitor therefore offers frequency-dependent opposition to the flow of alternating current. By means of calculus, it can be

---

* For additional information on capacitance, see *abc's of Capacitors* written by William F. Mullin and published by Howard W. Sams & Co., Inc., Indianapolis, Indiana 46268.

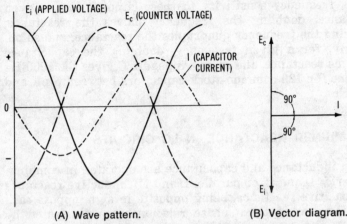

(A) Wave pattern.   (B) Vector diagram.

Fig. 1-13. Capacitor current/voltage relationship.

developed that this opposition is equal to the reciprocal of ωC. This opposition is termed *capacitive reactance*, is measured in ohms, and is given by:

$$X_c = \frac{1}{\omega C} = \frac{1}{2\pi f C} \qquad (1\text{-}18)$$

where,
 X_c is the capacitive reactance in ohms,
 C is the capacitance in farads,
 f is the frequency in hertz,
 ω is $2\pi f$,
 π is 3.1416.

The relations between current, voltage, and reactance are expressed in a form often called *Ohm's law for ac:*

$$X_c = \frac{E}{I}, \quad E = IX_c, \text{ and } \quad I = \frac{E}{X_c} \qquad (1\text{-}19)$$

where,
 X_c is in ohms,
 I is in amperes,
 E is in volts.

From Equation 1-19, it is seen that an alternating current flowing in a capacitive reactance produces a voltage drop IX_c; and because of the phase of capacitive reactance, this voltage lags the current by 90°.

Capacitive reactance is a major factor in all LC circuits used on ac. Like inductive reactance, it is the property that

25

imparts frequency sensitivity to these circuits. For a given capacitance, doubling the frequency halves the reactance, quartering the frequency quadruples the reactance, and so on. Similarly, for a given frequency, doubling the capacitance halves the reactance, and so on. Appendix C gives the 1000-Hz reactance for 126 common-stock capacitors between 5 pF and 5000 μF.

1.7 COMBINED REACTANCE IN LC CIRCUITS

When inductance and capacitance are operated in combination (series L and C, or parallel L and C), inductive reactance and capacitive reactance, being opposite in sign, oppose each other. Because of these phase relationships—illustrated by Figs. 1-6 and 1-10, respectively—the total reactance in the circuit is the difference between the two:

$$X_t = X_L - X_c \qquad (1\text{-}20)$$

where,

X_t, X_L, and X_c are in the same units (ohms, kilohms, megohms, etc.).

Illustrative Example: A 16-henry inductor and 1-microfarad capacitor are operated in series at 400 hertz. Calculate the total reactance of this combination.
From Equation 1-12, $X_L = \omega L = 40{,}212$ ohms; and from Equation 1-18, $X_c = 1/\omega C = 398$ ohms.
From Equation 1-20, $X_t = 40{,}212 - 398 = 39{,}814$ ohms.

1.8 RESONANCE

When inductance and capacitance are operated in combination, inductive reactance X_L is dominant at low frequencies and capacitive reactance X_c is dominant at high frequencies. At a sufficiently low frequency where $X_L \gg X_c$, the phase angle (θ) of the circuit will reach a maximum value of $+90°$; and at a sufficiently high frequency where $X_c \gg X_L$, the phase angle will reach a maximum value of $-90°$. At a frequency somewhere between these two extremes, which depends upon the L and C values, the total reactance, X_t, of the circuit is zero (since at that point $X_t = X_L - X_c = 0$), and the phase angle is zero. The frequency at which this situation occurs is the *resonant frequency* (f_r) of that particular LC combination:

$$f_r = \frac{1}{2\pi\sqrt{LC}} \qquad (1\text{-}21)$$

where,
  f is in hertz,
  L is in henrys,
  C is in farads,
  π is 3.1416.

Illustrative Example: Calculate the resonant frequency, in kHz, of a circuit containing 1 mH and 250 pF.
Here, 1 mH = 0.001 H, and 250 pF = 2.5×10^{-10} F.
From Equation 1-21 $f_r = 1/(6.2832\sqrt{0.001 \times 2.5 \times 10^{-10}}) =$
$1/(6.2832\sqrt{2.5 \times 10^{-13}}) = 1/(6.2832 \times 5 \times 10^{-7}) =$
$1/(3.1416 \times 10^{-6}) = 318{,}309$ Hz = 318.3 kHz.

1.9 FIGURE OF MERIT, Q

Since practical inductors and capacitors have inherent resistance, they can function efficiently as reactors only when this resistance is low. The ratio of reactance to resistance therefore is an indicator of this effectiveness and is termed *figure of merit*, symbolized by the letter Q. Thereby, $Q = X_L/R = X_c/R$, where X_L, X_c, and R are in ohms. The resistance acts for the most part in series with the inductor or capacitor. For the inductor:

$$Q = \frac{2\pi fL}{R} \qquad (1\text{-}22)$$

where,
  f is in hertz,
  L is in henrys,
  R is in ohms.

For the capacitor:

$$Q = \frac{1}{2\pi fCR} \qquad (1\text{-}23)$$

where,
  f is in hertz,
  C is in farads,
  R is in ohms.

The series resistance of a capacitor is virtually impossible to measure directly, and usually is determined by calculation from Q measurements ($R = X_c/Q$). The resistance of an inductor at dc and low frequencies is quite entirely the dc resistance of the wire in the coil. While at low frequencies, the resistance component has the same value as the easily measured dc resistance, at very high frequencies the resistance is a combination of dc resistance and all in-phase opposition arising from skin effect, influence of dielectrics in the magnetic field, and influence of shielding.

1.10 NATURE OF PRACTICAL INDUCTOR

It was mentioned earlier that a practical inductor, unlike the ideal pure inductance, has inherent resistance. It also has internal capacitance (termed *distributed capacitance*). These stray components are shown in relationship to the inductance, in Fig. 1-14. The resistance is occasioned by the wire with which the coil is wound and, at very high frequencies, also by skin effect and other factors. It is minimized by using thick wire or, especially at radio frequencies, braided wire. The shunting capacitance results from capacitor effect between adjacent turns and between layers of turns, and it is minimized by spacing the turns and by special styles of winding in which the turns are crisscrossed to destroy their adjacency.

Internal R and C_d combine with L to make an inductor an impedance, rather than a simple reactor, and in many applications it must be dealt with as such. The detrimental effect of resistance has already been pointed out in the discussion of Q in Section 1.9. Depending upon frequency, the distributed capacitance can limit the range over which a given inductor can resonate with a selected external capacitor.

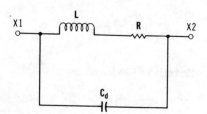

Fig. 1-14. Equivalent circuit of practical inductor.

1.11 PURE L AND C IN COMBINATION

For the moment, we will neglect the resistance that is always present, in however small amount, in LC circuits and will consider the effect of pure inductance in combination with pure capacitance, the ideal LC circuit. The effect of resistance will be saved for separate discussion in Section 1.12.

Inductance and Capacitance in Series

In the series LC circuit (see Fig. 1-15A), the total reactance is equal to $X_L - X_c$ and is zero at resonance. As the frequency is varied, the phase angle increases to $+90°$ at the frequency at which $X_L \gg X_c$, and increases to $-90°$ at the frequency at which $X_c \gg X_L$. The angle is zero at resonance, where $X_L = X_c$.

Inductance and Capacitance in Parallel

In the parallel LC circuit (see Fig. 1-15B), the total reactance is equal to $L[C(X_L - X_c)]$, and is infinite at resonance.

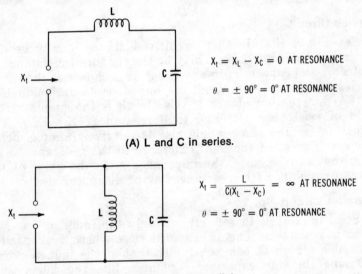

$X_t = X_L - X_c = 0$ AT RESONANCE

$\theta = \pm 90° = 0°$ AT RESONANCE

(A) L and C in series.

$X_t = \dfrac{L}{C(X_L - X_c)} = \infty$ AT RESONANCE

$\theta = \pm 90° = 0°$ AT RESONANCE

(B) L and C in parallel.

Fig. 1-15. Basic combinations of pure inductance and capacitance.

As the frequency is varied over a sufficient range, the phase angle is zero at resonance, and it increases to $+90°$ when $X_L \gg X_c$ and increases to $-90°$ when $X_c \gg X_L$.

As the losses in inductors and capacitors are reduced, performance approaches that of the ideal inductor and capacitor. Indeed, when extremely high-Q inductors and capacitors are available for a given application, their remaining resistance sometimes is so small as to be ignored so that circuit design may proceed as if pure inductance and capacitance were being used. In most instances, however, the resistive component cannot be ignored, and its influence is treated in the next section.

1.12 PRACTICAL L AND C IN COMBINATION

Since the inherent resistance in an LC circuit usually cannot be neglected, it must be dealt with in most analyses and design of LC circuits. At resonance, reactance disappears, but in a practical LC circuit resistance does not. An LC circuit thus is most often an LCR circuit; and when resistance enters, we must talk about impedance, not just simple reactance. Impedance $Z_{ohms} = \sqrt{R_{ohms}^2 + X_{ohms}^2}$. Fig. 1-16 shows the basic LCR arrangements.

Series Circuit

See Fig. 1-16A. In this circuit, resistance R may be due entirely to inductor L or it may be the combined resistance of inductor and capacitor. In any event, it is shown as the single series resistance. Note from the impedance equation in Fig. 1-16A that the impedance of this circuit is the simple vector sum of resistance (R) and total reactance $(X_L - X_c)$. The equation for the phase angle also is relatively simple, being the arc tangent of the total reactance to the resistance. Note also that at resonance, the impedance is R, the value of the circuit resistance (the reactance having disappeared).

Parallel Circuit No. 1

See Fig. 1-16B. In this circuit, only the inductor has significant resistance. This is often the case, where the capacitor has such a high Q, compared with that of the inductor, that entering the tiny capacitor resistance into the calculations will needlessly complicate them and contribute little to the final result. Note that the impedance of this circuit is more complicated than that of Fig. 1-16A.

Parallel Circuit No. 2

See Fig. 1-16C. While the condition shown in Fig. 1-16B is often the case in practice, it is not always so. There are many instances in which the inductor and capacitor each have significant resistance (R_L and R_c). Note here that the impedance equation and the phase-angle equation are the most complex of all.

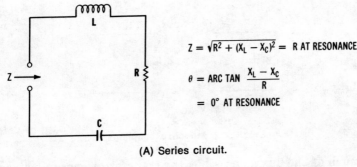

$$Z = \sqrt{R^2 + (X_L - X_C)^2} = R \text{ AT RESONANCE}$$

$$\theta = \text{ARC TAN } \frac{X_L - X_C}{R}$$

$$= 0° \text{ AT RESONANCE}$$

(A) Series circuit.

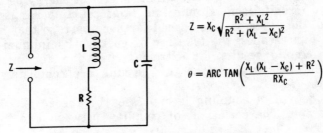

$$Z = X_C \sqrt{\frac{R^2 + X_L{}^2}{R^2 + (X_L - X_C)^2}}$$

$$\theta = \text{ARC TAN}\left(\frac{X_L (X_L - X_C) + R^2}{R X_C}\right)$$

(B) Parallel circuit (inductor resistive).

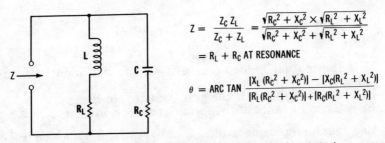

$$Z = \frac{Z_C Z_L}{Z_C + Z_L} = \frac{\sqrt{R_C{}^2 + X_C{}^2} \times \sqrt{R_L{}^2 + X_L{}^2}}{\sqrt{R_C{}^2 + X_C{}^2} + \sqrt{R_L{}^2 + X_L{}^2}}$$

$$= R_L + R_C \text{ AT RESONANCE}$$

$$\theta = \text{ARC TAN } \frac{[X_L (R_C{}^2 + X_C{}^2)] - [X_C(R_L{}^2 + X_L{}^2)]}{[R_L(R_C{}^2 + X_C{}^2)] + [R_C(R_L{}^2 + X_L{}^2)]}$$

(C) Parallel circuit (inductor and capacitor both resistive).

Fig. 1-16. Basic LCR circuits.

1.13 INDUCTIVE COUPLING

Two inductors are *coupled* when the magnetic field of one cuts the turns of the other so that energy is transmitted from one to the other or, conversely, so that energy is absorbed by one from the other. Thus, in Fig. 1-17, the magnetic flux resulting from current flow in the primary L1C1 circuit induces a voltage across inductor L2, and this causes a current to flow in the secondary, L2C2, circuit. If the inductors are placed as close together as possible and correctly oriented, so as to utilize

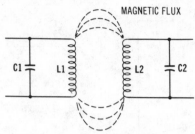

Fig. 1-17. Inductive coupling.

as much of the flux as possible, the transfer of energy between the two circuits is maximum and *tight coupling* results. If, instead, the inductors are spaced farther apart, so that much of the flux escapes, the transfer of energy is minimum and *loose coupling* results. At extreme separation (or in some cases 90° orientation), the two circuits are completely *decoupled*.

The degree of coupling is expressed by the *coefficient of coupling*, k. The coefficient of coupling between two inductors is:

$$k = \frac{M}{\sqrt{L1L2}} \qquad (1\text{-}24)$$

where,

k is the coefficient of coupling,
M is the mutual inductance between the two inductors, in henrys,
L1 is the inductance of the first inductor, in henrys,
L2 is the inductance of the second inductor, in henrys.

The maximum value which k can reach is 1, which corresponds to 100% coupling. This figure can result only if every line of force in the linking flux is utilized.

1.14 TIME CONSTANT

Capacitors and inductors both are subject to delayed response when they are operated in series with resistors. When a capacitor is charged through a resistor, time is required for the charge to be completed, i.e., for the voltage across the capacitor to equal the source voltage. Similarly, when a voltage is applied to an inductor in series with a resistor, time is required for the current flowing through the inductor to stabilize, i.e., to rise to its final, steady value.

Capacitor-Resistor Time Constant

A capacitor takes time to charge or discharge through a resistor. The time interval required to charge or discharge to a standard percentage of the final value is termed the *time constant* (T) of the RC circuit.

Numerically, the time constant of the RC combination is defined as the time required to charge the capacitor to a voltage equal to $1 - 1/\epsilon$ of the final, fully charged voltage (i.e., to 63.2% of final voltage) or to discharge the capacitor to a voltage equal to $1/\epsilon$ of the initial, fully charged voltage (i.e., to 36.79% of initial voltage). For practical purposes, these figures are usually rounded off to 63% and 37%, respectively.

The time constant of an RC circuit is calculated in the following manner:

$$T = RC \qquad\qquad (1\text{-}25)$$

where,

T is the time constant in seconds,
R is the resistance in ohms,
C is the capacitance in farads.

Thus, the time constant of an RC circuit containing 0.002 μF and 100,000 ohms is $T = 100,000\,(2 \times 10^{-9}) = 0.0002$ s $= 0.2$ ms.

For the same RC circuit, T has the same value for discharge as for charge. Fig. 1-18 illustrates the progress of charge (Fig. 1-18A) and of discharge (Fig. 1-18B).

The time-constant equation can be rewritten to determine either R or C for a desired value of time constant: $R = T/C$, and $C = T/R$. Thus, the resistance that must be used with an available 0.005-μF capacitor for a 1-microsecond time constant $= T/C = (1 \times 10^{-6})/(5 \times 10^{-9}) = 200$ ohms. Similarly, the capacitance required with a 50,000-ohm resistor for a time constant of 48 seconds $= T/R = 48/50,000 = 960$ μF.

Appendix D gives the time constants of a number of common RC combinations ranging from 0.0001 microfarad with 1 ohm to 1000 microfarads with 10 megohms.

Inductor-Resistor Time Constant

After application of voltage, a time interval is required for current to reach a maximum in an inductor operated in series with a resistor. Numerically, the inductor-resistor time constant is defined as the time required for the current flowing through the inductor to reach a value equal to $1 - 1/\epsilon$ of its final, steady value (i.e., to 63% of final value).

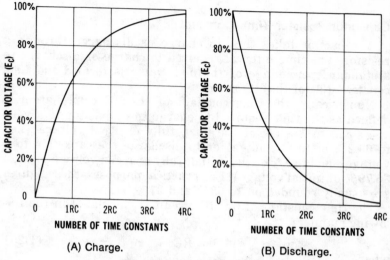

(A) Charge. (B) Discharge.

Fig. 1-18. RC time constant.

The time constant of an LR circuit is calculated in the following manner:

$$T = \frac{L}{R} \qquad (1\text{-}26)$$

where,

T is the time constant in seconds,
L is the inductance in henrys,
R is the resistance in ohms (including internal resistance of the inductor).

Thus, the time constant of an LR circuit containing a 20-henry inductor (internal resistance, 900 ohms) in series with 5000 ohms is $T = 20/(900 + 5000) = 20/5900 = 0.0034$ s $= 3.4$ ms.

Fig. 1-19 illustrates the progress of current growth in an LR circuit.

The time-constant equation can be rewritten to determine either R or L for a desired value of time constant: $L = RT$, and $R = L/T$. Thus, the total resistance (external resistance plus inductor-coil resistance) that must be used with an available 8-henry inductor for a 10-millisecond time constant is $R = L/T = 8/0.01 = 800$ ohms. Similarly, the inductance required with a 27-ohm resistor for a time constant of 2 seconds is $L = RT = 27(2) = 54$ H.

Appendix E gives the time constants of a number of inductor-resistor combinations ranging from 100 microhenrys with 1 ohm to 1000 henrys with 10 megohms.

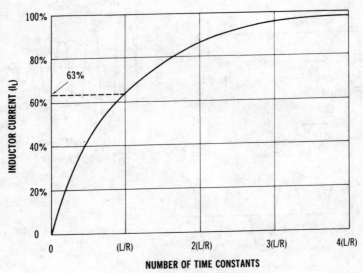

Fig. 1-19. L/R time constant.

1.15 OSCILLATIONS IN LC CIRCUIT

If a capacitor is charged (say, with its upper plate positive, as shown by the solid + in Fig. 1-20A) and then has an inductor, L, connected across it, the capacitor will proceed to discharge through the inductor, the electrons on the negative plate passing through the inductor, as shown by the solid arrow. This action, transferring the electrons from one plate to the other, recharges the capacitor to the opposite polarity, and the current causes a magnetic field to expand about the inductor. But when the capacitor becomes fully recharged, the current ceases and the magnetic field collapses. The collapsing field then induces a voltage of opposite polarity at the inductor terminals, and this voltage discharges the capacitor, causing a current to flow in the opposite direction (see dotted arrow) and the original polarity of the capacitor to be restored. The action then repeats itself periodically as the capacitor charges in first one direction and then the other, and the inductor field alternately expands and collapses. An alternating current thus flows in the circuit.

Once started, this action would continue forever, but the circuit has internal resistance (see dotted R in Fig. 1-20A) which absorbs energy and eventually stifles the oscillations. Thus, when a single impulse starts the chain of events, the

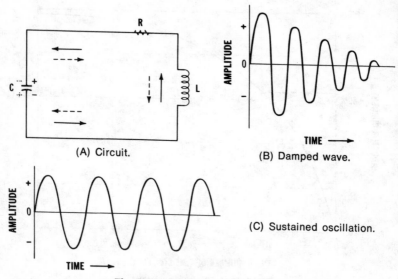

(A) Circuit.

(B) Damped wave.

(C) Sustained oscillation.

Fig. 1-20. Oscillations in LC circuit.

amplitude of each cycle, like the swing of a pendulum given only a single push, will be lower than that of the preceding one (see Fig. 1-20B) until the oscillations finally die out. A *damped wave* is the result. The higher its Q, the longer the circuit will "ring" in response to a single exciting pulse.

In a complete practical oscillator circuit, a tube or transistor supplies just enough energy in the proper phase to overcome the losses of the LC circuit, and this sustains the oscillations at constant amplitude to correct the damping shown in Fig. 1-20B. Thus, the LC circuit actually is the oscillating medium, but its action must be sustained, like that of a pendulum or flywheel, by little pushes of energy from a tube or transistor.

1.16 RANGE OF APPLICATION OF LC CIRCUITS

Both fixed and variable LC circuits enjoy a wide range of applications. A few of the common devices in which these circuits provide the basis of operation are af and rf tuning networks, tuned transformers, filters (both signal and power supply), voltage regulators, discriminators, ratio detectors, wavetraps, and test instruments (bridges, comparators, wavemeters, null devices, etc.).

CHAPTER 2

Tuned Circuits

The first business of the LC circuit was the tuning of radio equipment; and, after more than three-quarters of a century, this simple circuit is still doing that job—with the added function of tuning television. It is doing other jobs, both at radio and audio frequencies; however, even in its most sophisticated modern applications far removed from simple tuning, the LC circuit still is, at core, providing selectivity.

Advancing the explanation introduced in Chapter 1, the present chapter describes the LC circuit *qua* tuner, and presents a number of its applications in that function. There are many more, of course, than can be included in this chapter. But even this brief survey should highlight the versatility of this simple combination of passive components.

The reader should return often to Sections 1.11 and 1.12 in Chapter 1 if he needs to strengthen his understanding of the material in the present chapter.

2.1 SERIES-RESONANT CIRCUIT

Fig. 2-1A shows a series-resonant circuit, so called because in this arrangement, generator GEN, inductance L, capacitance C, and internal resistance R are connected in series. The resonant frequency (f_r) of the circuit depends upon the L and C values in the following relationship:

$$f_r = \frac{1}{2\pi\sqrt{LC}} \qquad\qquad (2\text{-}1)$$

where,

f$_r$ is the resonant frequency in hertz,
L is the inductance in henrys,
C is the capacitance in farads,
π is 3.1416.

Illustrative Example: Calculate the resonant frequency in megahertz of a series-resonant circuit containing an inductance of 50 μH and a capacitance of 75 pF.
Here, $L = 5 \times 10^{-5}$ H, and $C = 7.5 \times 10^{-11}$ F.
From Equation 2-1, $f_r = 1/6.283 \sqrt{5 \times 10^{-5} (7.5 \times 10^{-11})} = 1/6.283$ $\sqrt{3.75 \times 10^{-15}} = 1/6.283 \ (6.124 \times 10^{-8}) = 1/3.845 \times 10^{-7} =$ 259,895 Hz = 2.599 MHz.

When the inductance and capacitance are fixed, the circuit has a single resonant frequency. Often, however, either L or C or both are variable, and the circuit can be adjusted over a range of resonant frequencies.

In Fig. 2-1A, the ac generator delivers a constant voltage (E), forcing a current (I) through the circuit. At resonance, the inductive reactance and the capacitive reactance cancel each other, leaving only the internal resistance to determine the current. Maximum current therefore flows at resonance. (This is another way of saying that the impedance of the series-resonant tuner is lowest at resonance.) This performance is shown in Fig. 2-1B. Note from these curves that the sharpest response and greatest height (highest selectivity) are obtained when resistance is lowest, and the broadest response and lowest height when resistance is highest. High selectivity (sharp tuning) corresponds to high Q, and vice versa. The response in Fig. 2-1B may be obtained by either holding L and C constant and varying the generator frequency, or by holding the frequency constant and varying L or C. It is interesting to note that to the generator the series-resonant circuit looks like an inductor (in series with the internal resistance of the latter) at frequencies above resonance, like a capacitor (in series with the internal resistance) at frequencies below resonance, and like a resistor (the internal resistance) at resonance.

An important effect observed in the series-resonant circuit is *resonant voltage step-up.* This means that at resonance, the voltage (E$_L$) across the inductor and the voltage (E$_c$) across the capacitor each will be higher than the generator voltage. However, this phenomenon does not violate Kirchhoff's second

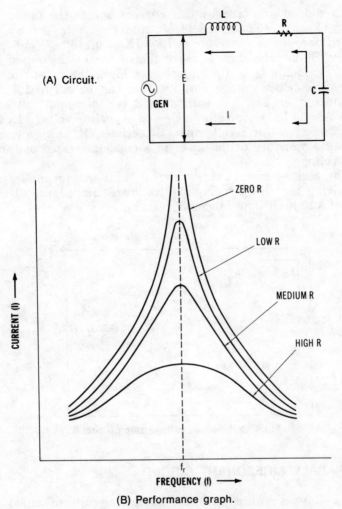

(A) Circuit.

(B) Performance graph.

Fig. 2-1. Series-resonant circuit.

law, for E_L and E_c are 180° out of phase with each other, and the sum of the voltage drops around the circuit equals the supply voltage. Fig. 2-2 illustrates this effect. In this arrangement, the input voltage (E_i) is 1 V at 1000 Hz, the resonant frequency of the LC combination. Thus, $X_L = X_c = 62.83$ ohms, and internal resistance R = 10 ohms. Since the total reactance at resonance is zero, the current flowing in the circuit depends entirely upon the resistance; i.e., I = E/R = 1/10 = 0.1 A, and this current flows through the inductor and capacitor. The

39

voltage drop produced by this current across the inductor is $E_L = IX_L = 0.1(62.83) = 6.28$ V; and the voltage drop produced across the capacitor is $E_c = IX_c = 0.1(62.83) = 6.28$ V. Thus, E_L and E_c each is more than 6 times the input voltage, E_i. If the Q of the inductor is high, the extent of this transformerless voltage amplification can be surprising. For instance, if in Fig. 2-2, resistance R is 1 ohm, current I will be 1 A, and inductor voltage E_L and capacitor voltage E_c each will be 62.8 V for 1-volt input. Sometimes, E_c is high enough to cause puncture or flashover in a capacitor rated at the input voltage.

The series-resonant tuner circuit finds use in applications requiring that the circuit have its lowest impedance at resonance and high impedance off resonance.

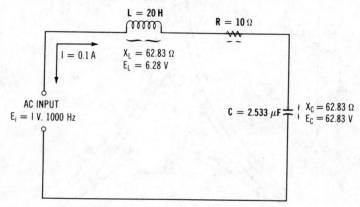

Fig. 2-2. Resonant voltage step-up circuit.

2.2 PARALLEL-RESONANT CIRCUIT

Fig. 2-3A shows a parallel-resonant circuit, so called because in this arrangement, generator GEN, inductance L, and capacitance C are connected in parallel. The internal resistance of this circuit is mainly in the inductor leg, as shown by R. As in the series-resonant circuit, the resonant frequency (f_r) of the parallel-resonant circuit depends upon the L and C values, in the same relationship. Therefore, Equation 2-1 applies also to the parallel circuit.

When the inductance and capacitance are fixed, the circuit has a single resonant frequency. Often, however, either L or C or both are variable, and the circuit can be adjusted over a range of resonant frequencies.

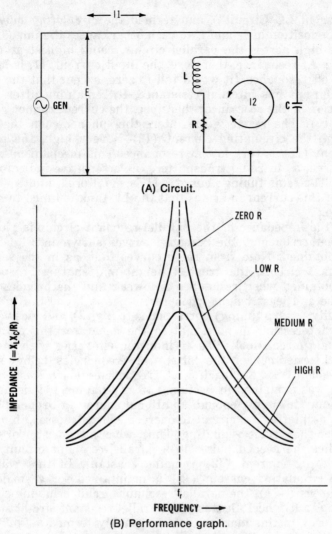

(A) Circuit.

(B) Performance graph.

Fig. 2-3. Parallel-resonant circuit.

In Fig. 2-3A, the ac generator delivers a constant voltage (E), forcing a current (I1) through the parallel circuit. Within the circuit, this current divides into two components (I_L in the inductive leg and I_c in the capacitive leg—neither shown) which are out of phase with each other. The net result of these two currents is current I2 flowing within the LC circuit as a result of charge and discharge of the capacitor through the inductor (See Section 1.15 in Chapter 1, "Oscil-

41

lations in LC Circuit") and termed the *circulating current*. The opposition of I_L and I_c to each other causes the impedance to be high across the parallel circuit, being highest at resonance. At resonance, therefore, the input current, I1, is lower than off resonance (it would fall to zero, except that the parallel circuit has internal resistance, R, remaining after cancellation of the reactance; therefore, the current cannot reach infinity). The result is the interesting phenomenon that although the circulating current (I2) is very high, the input current (I1) is very low at resonance. This mechanism is often used, as in radio transmitters, as a sensitive indicator of parallel-circuit tuning: current I1 is monitored, and a sharp dip of this current indicates that the LC tank is tuned to resonance.

If the impedance of the parallel-resonant circuit is plotted against frequency, the response curves shown in Fig. 2-3B are obtained. Note from these curves that, as in the series-resonant circuit, the greatest height and sharpest response are obtained when resistance is lowest, and the broadest response and least height when resistance is highest. High selectivity (sharp tuning) corresponds to high Q, and vice versa. It is interesting to note that to the generator the parallel-resonant circuit looks like an inductor (in series with the internal resistance of the latter) at frequencies below resonance, and like a capacitor at frequencies above resonance.

As was explained in Section 1.15 in Chapter 1, it is necessary only to supply enough additional energy in proper phase to an oscillating LC circuit to overcome circuit losses, in order to keep the circuit oscillating. Thus, when a transistor or tube supplies the needed pulses to a parallel-resonant circuit, the circulating current (I2) remains constant. In this way, a large circulating current (I2) is maintained for a small input current (I1), the parallel-resonant circuit behaving somewhat as a flywheel. Because the parallel-resonant circuit stores ac energy in this manner, it is frequently termed a *tank circuit*, or just a *tank*.

The parallel-resonant tuner circuit finds use in applications requiring that the tuned circuit have its highest impedance at resonance and low impedance at frequencies off resonance.

2.3 RESONANT-CIRCUIT CONSTANTS

The resonant frequency of a tuned circuit can be determined from Equation 2-1 when L and C both are known. In some instances, either L or C will be unknown, and the equation may be rewritten for that component:

Inductance unknown

$$L = \frac{1}{4\pi^2 f^2 C} \qquad (2\text{-}2)$$

where,
    L is the unknown inductance in henrys,
    C is the available capacitance in farads,
    f is the desired resonant frequency in hertz,
    $4\pi^2$ is 39.5.

Illustrative Example: An accurate 0.0047-μF capacitor is available. What value of inductance (in mH) will resonate this capacitance to a frequency of 1 MHz?
Here, C = 4.7 $\times$ 10^{-9}F, and f = 1 $\times$ 10^6 Hz.
From Equation 2-2, L = 1/(39.5 (1 $\times$ 10^6)^2 4.7 $\times$ 10^{-9}) = 1/(39.5 (1 $\times$ 10^{12}) 4.7 $\times$ 10^{-9}) = 1/30.5 (4.7 $\times$ 10^3) = 1/185,650 = 5.39 $\times$ 10^{-6}H = 0.00539 mH.

Capacitance unknown

$$\frac{1}{4\pi^2 f^2 L} \qquad (2\text{-}3)$$

where,
    C is the unknown capacitance in farads,
    L is the available inductance in henrys,
    f is the desired resonant frequency in hertz,
    $4\pi^2$ is 39.5.

Illustrative Example: An 8.5-henry inductor is available. What value of capacitance (in μF) will resonate this inductance to a frequency of 120 Hz?
From Equation 2-3, C = 1/(39.5(120^2) 8.5) = 1/(39.5(14,400)8.5) = 1/4,834,800 = 2.068 $\times$ 10^{-7}F = 0.2068 μF. (This value can be made up by connecting one 0.2- and one 0.0068-μF capacitor in parallel.)

Appendix F gives the resonant frequency of a number of combinations from 1 microhenry and 10 picofarads to 1000 henrys and 1000 microfarads, and applies both to series-resonant and parallel-resonant circuits.

2.4 SELECTIVITY

Selectivity is the sharpness of response of a tuned circuit. It expresses the ability of a pass-type circuit to select a signal of desired frequency and suppress all others, or the ability of a reject-type circuit to suppress a signal of undesired frequency and pass all others.

Fig. 2-4 shows a selectivity curve. A plot of this type is obtained by applying a constant input voltage to the tuned circuit and tuning the generator above and below resonance. The resonant point is indicated by peak upswing (as in Fig. 2-4) of voltage or current in a pass-type circuit, or by maximum dip (where Fig. 2-4 would be inverted) in a reject-type circuit. Voltage, current, or arbitrary units may be plotted along the vertical axis, frequency along the horizontal. The 100% point in this illustration merely designates maximum height (current, voltage, etc.) of the curve under study. The selectivity curve consists of its tip (called the *nose*) and its sides (called the *skirts*). Depending upon the characteristics of the circuit, some curves are blunt nosed and others are pointed.

It is apparent from the width of the curve that a practical tuned circuit is not completely single-frequency responsive, but operates over a narrow band of frequencies. The more selective the circuit, the narrower this band. At any point, the *bandwidth* of the curve—in Hz, kHz, or MHz—is the width at a selected point. This point is usually stated. Thus, in Fig. 2-4, the bandwidth at 70.7% of maximum rise (i.e., 3 dB down

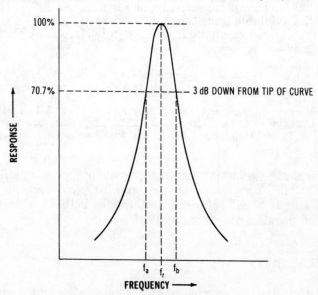

Fig. 2-4. Selectivity curve.

from maximum) is equal to the frequency difference $f_b - f_a$. A sharply tuned circuit has good *skirt selectivity* (the ratio of the bandwidth at one point, say 3 dB, to that at another point, say 60 dB).

2.5 CIRCUIT Q

It is clear from the resonance curves shown earlier—Figs. 2-1B and 2-3B—that high selectivity is the result of high Q, and vice versa. This permits the value of circuit Q to be approximated closely by finding the ratio of resonant frequency (f_r) to bandwidth (BW). This is done in the laboratory by inductively coupling the tuned circuit loosely to a constant-output-voltage signal generator, tuning the generator above and below the resonant frequency, noting the current or voltage response, and calculating Q as the ratio of resonant frequency to bandwidth at the −3-dB point: (1) With reference to Fig. 2-4, tune the generator to the resonant frequency (f_r) and note the value of circuit voltage or current at this point. (2) Detune the generator below resonance to the frequency (f_a) at which the circuit voltage or current falls to 70.7% of its resonant value. This is 3 dB below resonant voltage or current. (3) Next, detune the generator above resonance to the frequency (f_b) at which the circuit voltage or current falls again to 70.7%. (4) Then, calculate Q:

$$Q = \frac{f_r}{BW} = \frac{f_r}{\Delta f} = \frac{f_r}{f_b - f_a} \qquad (2\text{-}4)$$

where,
   all frequencies (f's) are in the same unit (Hz, kHz, or
   MHz).

Illustrative Example: A certain tuned circuit is tested with a signal generator and high-input-impedance electronic ac voltmeter. The resonant frequency (f_r) is found to be 2.5 MHz, and the generator output voltage is adjusted for exactly 100 mV across the inductor. The generator then is detuned below resonance to the frequency at which the inductor voltage falls to 71 mV (approximately 70.7% of resonant voltage), and this frequency (f_a) is 2.3 MHz. Next, the generator is detuned above resonance to the frequency at which the inductor voltage again falls to 71 mV, and this frequency (f_b) is 2.7 MHz. Calculate the Q of this circuit.

From Equation 2-4, $Q = \dfrac{2.5}{2.7 - 2.3} = \dfrac{2.5}{0.4} = 6.25$.

Drawing energy from a tuned circuit ("loading" the circuit) tends to reduce the selectivity and Q, because the drain constitutes a loss which resembles the equivalent resistance reflected back into the tuned circuit. This action is reduced by lightly loading the circuit whenever possible.

2.6 COUPLED RESONANT CIRCUITS

When identical resonant circuits are cascaded, the selectivity of the combination is higher than that of any one of the circuits. Thus, the tuning of an electronic equipment is sharpened (bandwidth reduced) by cascading several LC circuits for ganged tuning. Often, an amplifying device—tube, transistor, or IC—operates between successive LC circuits, but in many instances the resonant circuits are simply capacitance coupled, as shown in Fig. 2-5A. Paradoxically, coupled resonant circuits are used also for the opposite purpose: to broaden tuning under controlled conditions. For example, one of two circuits is resonated at a higher frequency than the other, and the two circuits together offer a wider passband (broader-nosed curve) than is afforded by one circuit alone.

In Fig. 2-5A, the resonant circuits (L1,C1, L2,C3, L3C5) are spaced far enough apart to eliminate mutual inductance between them, and the coupling of energy from one to the other is entirely via capacitors C2 and C4. Although this circuit narrows the passband, the resonant voltage may be reduced across successive tanks, depending upon the reactance of C2 and C4.

Inductive coupling (also called *transformer coupling*) takes many forms. In every case, however, the coupling is accomplished by means of the mutual inductance between the two inductors. The input is called the *primary* circuit, and the output the *secondary*.

Fig. 2-5 (B to E) shows several examples. In Fig. 2-5B, the input and output circuit each is series resonant; in Fig. 2-5C, input is series resonant and output is parallel resonant; in Fig. 2-5D, input is parallel resonant and output is series resonant; and in Fig. 2-5E, input and output each is parallel resonant. Series resonance is used for low-impedance input or output, and parallel resonance for high-impedance input or output. The parallel-in/parallel-out circuit is a familiar one, being the type found in rf, if, and detector transformers and in tuned audio transformers.

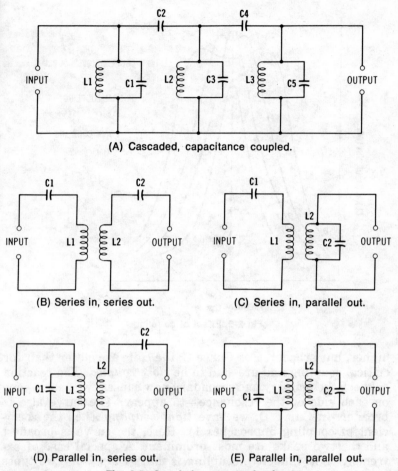

(A) Cascaded, capacitance coupled.

(B) Series in, series out.

(C) Series in, parallel out.

(D) Parallel in, series out.

(E) Parallel in, parallel out.

Fig. 2-5. Coupled resonant circuits.

In inductively coupled circuits, the degree of coupling greatly influences the selectivity of the circuit. The secondary loads the primary, and the primary loads the secondary; and this mutual action increases as the coupling grows tighter, tending to broaden the response. Fig. 2-6 shows the differences which result when a constant voltage is applied to the tuned circuit and the coupling is changed. The primary and secondary are tuned to the same frequency. In this illustration, Curve A—the sharpest—corresponds to loose coupling (coils well separated). Curve B shows *critical coupling*, the condition in which maximum energy is drawn from the tuned circuit. Here, the coils are more closely spaced, so the peak is

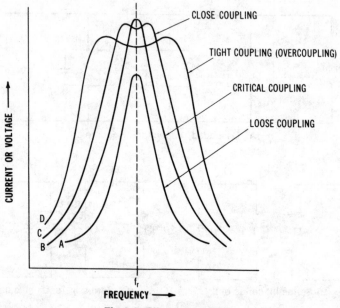

CLOSE COUPLING

TIGHT COUPLING (OVERCOUPLING)

CRITICAL COUPLING

LOOSE COUPLING

CURRENT OR VOLTAGE

D
C
B   A

f_r

FREQUENCY

Fig 2-6. Effect of coupling.

higher, but broader. For Curve C, the coils are closer than for critical coupling and are said to be *close coupled*. The reaction between the primary and secondary now causes two peaks—one on each side of the resonance—to appear. The curve also is broader. In Curve D, we have *tight coupling* (i.e., the coefficient of coupling approaches 1). Here, the curve is broadest and the two peaks are most prominent. A special case of extremely tight inductive coupling is *unity coupling*. Two types are shown in Fig. 2-7. As shown, the turns of secondary coil L2 (dotted line) are interwound with those of primary coil L1 (solid line). In Fig. 2-7B, primary coil L1 is wound with hollow tubing, and secondary coil L2 (dotted line) consists of an insulated wire threaded through this tubing coil. In unity coupling, mutual inductance M is so high that a single capacitor, C1, tunes both L1 and L2.

Fig. 2-8 shows several additional methods of coupling. Fig. 2-8A shows the familiar *link coupling*, which is widely used in radio transmitters and other equipment in which the two tuned circuits must be spaced far apart. The "links" consist of coils L2 and L3, each usually 1 to 3 turns wound around the "cold" end of the tank coil (L1, L4). They provide the inductive coupling which is impossible to obtain directly be-

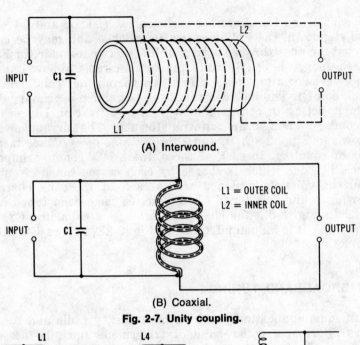

(A) Interwound.

L1 = OUTER COIL
L2 = INNER COIL

(B) Coaxial.

Fig. 2-7. Unity coupling.

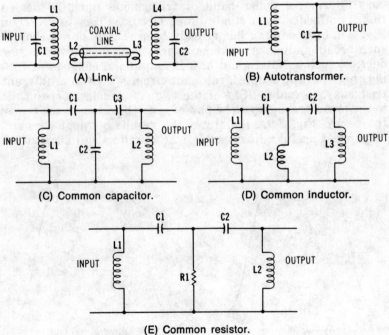

(A) Link.

(B) Autotransformer.

(C) Common capacitor.

(D) Common inductor.

(E) Common resistor.

Fig. 2-8. Miscellaneous coupling methods.

tween L1 and L4. Since the turns ratios (L1:L2 and L4:L3) are stepdown, the links are low impedance and may be connected together through a coaxial line or twisted pair. In Fig. 2-8B, energy is coupled into the tuned circuit at low impedance through a tap taken a few turns from the "cold" end of tank coil L1. The included turns thus establish a primary coil which, acting in conjunction with the entire coil, L1, as the secondary creates an autotransformer. The arrangements shown in Figs. 2-8C, D, and E are *common-impedance* methods of coupling. In each of these, there is a common impedance, i.e., one which is shared by each of the tuned circuits, and the voltage developed across each of these by current flowing in the first tuned circuit excites the second tuned circuit. In Fig. 2-8C, the shared impedance is capacitor C2, in Fig. 2-8D, it is inductor L2, and in Fig. 2-8E, it is resistor R1. tor R1.

2.7 BROADBAND TUNING

In some applications, such as high-fidelity radio and bandpass filtering, a wider band of frequencies must be passed than is afforded by a single tuned circuit. From Curve D in Fig. 2-6, overcoupling has already been shown as a means of such broadbanding. A more satisfactory method in fixed-tune devices, such as filters and hi-fi if amplifiers, consists of tuning the separate, coupled, resonant circuits each to a different frequency, f_{r1} and f_{r2}. One of the two frequencies corresponds to the lower frequency to be passed, and the other to the upper frequency. Fig. 2-9A illustrates the resulting "double-hump" response. Careful adjustment of the coupling will smooth the

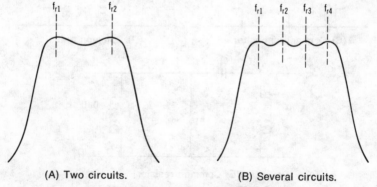

(A) Two circuits.   (B) Several circuits.

Fig. 2-9. Broadband response.

humps to some extent and give a flatter, broader-nosed, curve. For a wider band, several tuned circuits may be employed, being connected as shown in Fig. 2-5A, either with or without interposed amplifiers. The separate resonant frequencies result in somewhat of a ripple in the passband, as shown in Fig. 2-9B.

2.8 RANGE COVERAGE

There are many ways of determining the operating range of a tuned circuit and of presetting the circuit (aligning it) at the upper and lower frequency limits of that range. Fig. 2-10 shows a few of these methods.

In Fig. 2-10A, a fixed inductor and variable capacitor are employed. The capacitor has a minimum capacitance (C_{min}), as well as maximum capacitance (C_{max}), and these two determine the tuning range. From Equation 2-1, the upper frequency (f_{max}) must be calculated, using C_{min}; then, the lower

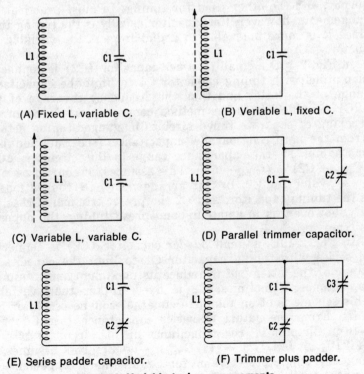

(A) Fixed L, variable C.

(B) Veriable L, fixed C.

(C) Variable L, variable C.

(D) Parallel trimmer capacitor.

(E) Series padder capacitor.

(F) Trimmer plus padder.

Fig. 2-10. Variable tuning arrangements.

frequency (f_{min}) must be calculated, using C_{max}. The tuning range, Δf, then is $f_{max} - f_{min}$, and the frequency coverage is $f_{min} + \Delta f$.

In Fig. 2-10B, inductor L1 is variable, and capacitor C1 is fixed. Inductance may be varied in several ways. In low-power rf applications and in some af applications, the inductance is adjusted by means of a slug, as indicated in the schematic; in transmitters and industrial oscillators, the coil may have a series of taps or a turns-contacting slider for this variation, or the coil can be a variometer (a device consisting of two coils in series, one rotatable inside the other). In Fig. 2-10B, the coil has a minimum inductance (L_{min}), as well as a maximum inductance (L_{max}) which will determine the tuning range. From Equation 2-1, the upper frequency (f_{max}) must be calculated, using L_{min}; then, the lower frequency (f_{min}) must be calculated, using L_{max}. The tuning range, Δf, then is $f_{max} - f_{min}$, and the frequency coverage is $f_{min} + \Delta f$.

In Fig. 2-10C, inductance and capacitance both are variable. Either one may be used to set the circuit to a range-limit frequency, and the other used for tuning. In most practical arrangements, however, the capacitor usually is the tuning unit, since it is more amenable to a dial than is the variable inductor.

In Fig. 2-10D a small *trimmer* capacitor (C2) is connected in parallel with tuning capacitor C1 to limit the capacitance range of the latter and thus the frequency coverage of the L1C1 circuit. This is the method commonly employed in the tracking of separate tuned circuits in ganged-tuning setups. From the nature of parallel capacitance (see Equation 1-17 in Chapter 1), the capacitance range in this circuit is $\Delta C = (C_{1max} + C2) - (C_{1min} + C2)$. This assumes, of course, a single preset value for C2. In this arrangement, C1 conventionally is the tuning capacitor, and C2 the preset trimmer. Occasionally, however, as in amateur bandspread tuning, the opposite is true.

In Fig. 2-10E, a small *padder* capacitor (C2) is connected in series with tuning capacitor C1 to limit the capacitance range of the latter *and* to reduce its maximum and minimum capacitance. Sometimes, C2 is fixed, rather than variable. This scheme is often found in superheterodyne oscillator circuits. From the nature of series capacitance (see Equation 1-16 in Chapter 1), the capacitance in this circuit is $\Delta C = 1/[(1/(C_{1max}) - 1/C2] - [(1/C_{1min}) - 1/C2]$. This assumes, of course, a single preset value for C2. In this arrangement, C1 conventionally is the tuning capacitor, and C2 the preset padder. Occasionally, however, as in amateur bandspread tuning, the opposite is true.

In Fig. 2-10F, a combination of the two preceding methods is employed. Both a series padder (C2) and parallel trimmer (C3) are employed for close pruning of the capacitance range of main tuning capacitor C1.

In each of the foregoing examples, the distributed capacitance (C_d) of the inductor has been neglected, since it is usually small compared to the capacitance of the tuning, trimmer, and padder capacitances in the circuit. But, in some radio-frequency circuits—especially where very small tuning capacitances are employed—distributed capacitance cannot be ignored. For example, a popular 10-mH inductor has a distributed capacitance of 4 pF. If this coil is used in a tuned circuit, such as Fig. 2-10A, with a 100-pF variable capacitor whose minimum capacitance is 11.5 pF, the capacitance in the circuit when the variable capacitor is set to "zero" is $C = C_{min} + C_d = 11.5 + 4 = 15.5$ pF. Residual capacitance such as this is very important in determining the upper frequency limit of a tuning range.

2.9 SELF-RESONANCE

Since the distributed capacitance (C_d) of an inductor acts in parallel with the inductance, a simple LC circuit is set up by this combination. Every inductor therefore is resonant at some frequency, its *self-resonant frequency*.

A search of manufacturers' ratings on inductors shows that these units have self-resonant frequencies ranging from 138 kHz to 690 MHz, depending upon inductance and type of winding. When designing a tuned circuit, one selects an inductor whose self-resonant frequency is as far as possible from the intended operating range of the circuit.

2.10 SYMMETRICAL CIRCUITS

The tuned circuits shown up to this point are all single ended. However, a respectable amount of electronic equipment is double ended, i.e., requiring balanced input or output (or both) tuned circuits. These arrangements are known also as *balanced, pushpull,* or *full-wave.*

Fig. 2-11 shows two forms of the symmetrical circuit. In each of these, inductor L1 is divided into two identical halves, the common point (center tap) usually being grounded. The voltage between COMMON and A is 180° out of phase with the voltage between COMMON and B. In Fig. 2-11A, each half of the inductor is tuned by an identical capacitor—C1 for the upper half, C2 for the lower half. When the circuit must be continuously tunable, C1 and C2 are the halves of a split-stator variable capacitor, so that the same capacitance will be offered for each at all settings. The L1C1 half of the circuit thus is identical to the L2C2 half. That is, the upper half of inductance L1 resonates with capacitance C1 to the same frequency as the lower half of the inductance with capacitance C2. For this reason, L1 is twice the size it would be in a single-ended circuit for the same frequency. In Fig. 2-11B, a single capacitance (C1) tunes the center-tapped inductor (L1), and the resonant frequency is calculated for the entire coil. This latter arrangement is not suitable in all applications requiring a variable capacitor, since the frame of the capacitor may exhibit unequal capacitance from A to COMMON and from B to COMMON.

The coupling methods shown in Figs. 2-5E and 2-8A may be employed with the symmetrical tanks. When link coupling is employed, the link coil is wound around the center of the tapped coil.

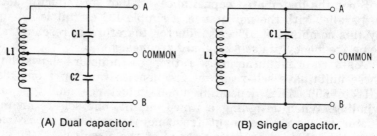

(A) Dual capacitor. (B) Single capacitor.

Fig. 2-11. Symmetrical-tuned circuits.

2-11 DC-TUNED CIRCUITS

Certain LC circuits can be tuned by means of an adjustable dc voltage. This is convenient in remote tuning, remote control of apparatus, automatic frequency control (afc), and various automation processes and telemetering. Fig. 2-12 shows two methods.

A *saturable reactor* (T₁) is employed in Fig. 2-12A and B. In this transformerlike device, dc flowing through the pri-

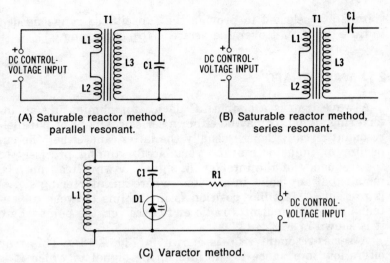

(A) Saturable reactor method, parallel resonant.

(B) Saturable reactor method, series resonant.

(C) Varactor method.

Fig. 2-12. Tuned, dc controlled circuits.

mary winding (L1 and L2 in series) changes the permeability of the core and thereby changes the inductance of the secondary winding (L3). The resonant circuit is comprised of secondary coil L3 and capacitor C1. The secondary and capacitor are connected for parallel resonance in Fig. 2-12A, and for series resonance in Fig. 2-12B. Primary coils L_1 and L_2 are connected in series bucking, so that no ac from the secondary is fed back to the dc control-voltage source. The sensitivity of the circuit is such that a small direct current flowing through the primary winding can control a large alternating current in the L3C1 tank. Special core materials afford operation at high frequencies.

A *varactor* (voltage-variable capacitor), D1, is employed in Fig. 2-12C. This is a specially processed silicon diode which is operated in the reverse-biased mode (anode negative, cathode positive). In this circuit, the resonant frequency is determined by the inductance of coil L1 and the dc-controlled capacitance of the varactor. Capacitor C1 is a blocking capacitor which prevents the coil from short-circuiting the dc control voltage; its capacitance is chosen much higher than that of the varactor (usually 1000 ×) for lowest reactance, so that the varactor—rather than capacitor C1—tunes the circuit. Isolating resistance R1 is very high (usually 1 to 10 megohms); and since the varactor is essentially voltage operated, there is virtually no current through this resistor and accordingly no voltage drop across it. The varactor and dc control voltage are selected to provide the capacitance range needed to tune the circuit over the desired frequency range.

2.12 WAVE TRAPS

A wave trap is a simple LC device for eliminating an interfering signal. It may be either a series-resonant or parallel-resonant circuit, though usually the latter, and either the capacitor or inductor may be variable for tuning precisely to the frequency of the interfering signal. A wave trap may be installed at any point in a system where an interfering signal is present. A familiar position is in the line between an antenna and the input of a radio or tv receiver, the point where it is shown in Fig. 2-13.

A series-resonant trap is shown in Fig. 2-13A. This trap offers low impedance to the interfering signal to which it is tuned, and the signal accordingly is shunted around the receiver to ground. Signals of other frequencies see the trap as a higher impedance, and pass by it to the receiver, with

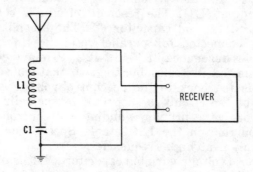

(A) Series-resonant type.

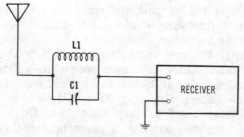

(B) Parallel-resonant type.

Fig. 2-13. Wave traps.

little loss. A parallel-resonant trap is shown in Fig. 2-13B. This trap offers high impedance to the interfering signal which it captures as a circulating current flowing inside the L1C1 tank. Removal of the signal corresponds to reduction of the line current, which is characteristic of the parallel-resonant circuit (see Section 2.2). Signals at other frequencies see the trap as a lower impedance, and pass through to the receiver with little loss. In either type of trap, the circuit Q must be as high as practicable, to minimize attenuation of desired signals. Another familiar position for one or more wavetraps is the output circuit of a transmitter, where the trap removes undesired harmonics of the radiated signal.

2.13 WAVEMETERS

Another familiar application of the simple LC tuned circuit is the *absorption wavemeter*, so called because its operation depends upon its absorption of a small amount of rf energy from the circuit to which it is inductively coupled. This instrument is also called an *absorption frequency meter*.

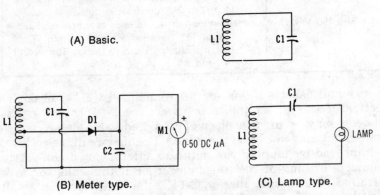

(A) Basic.

(B) Meter type.

(C) Lamp type.

Fig. 2-14. Wavemeters.

Fig. 2-14 shows three common versions of this instrument. Basically, it, like the wave trap, is a single-tuned circuit consisting of a fixed-inductance coil (L1) and variable capacitor (C1). The coil is loosely coupled to the tank of the oscillator or amplifier under test, by holding it near the latter, and the wavemeter is tuned to resonance at the unknown frequency (f_x) by adjusting the capacitor. The unknown frequency then is read from the calibrated dial of the capacitor. The frequency range (band) is changed by plugging in a different coil.

Resonance can be indicated in several ways. When the circuit in Fig. 2-14A is used, the unknown-signal source must have a current meter in its output stage (a plate milliammeter in a tube-type source, a collector milliammeter in a transistor-type source), and the deflection of this meter will rise sharply as the wavemeter is tuned, the peak point of this rise indicating resonance. In the two other arrangements—Figs. 2-14B and C—the wavemeter has a self-contained indicator. In Fig. 2-14B, a germanium diode (D1) rectifies the picked-up rf

Table 2-1. Wavemeter Coil Data

	Variable Capacitor C_1 = 140 pF
(A) 1.1—3.8 MHz	72 turns No. 32 enameled wire closewound on 1"-diameter plug-in form. Tap 18th turn from bottom.
(B) 3.7—12.5 MHz	21 turns No. 22 enameled wire closewound on 1"-diameter plug-in form. Tap 7th turn from bottom.
(C) 12—39 MHz	6 turns No. 22 enameled wire on 1"-diameter plug-in form. Space to winding length of ⅜ inch. Tap 3rd turn from bottom.
(D) 37—150 MHz	Hairpin loop of No. 16 bare copper wire. Spacing of ½" between legs of hairpin. Total length including bend: 2 inches Tap center of bend.

energy and deflects a 0-50–dc microammeter (M1), peak deflection of this meter indicating resonance. (A thermogalvanometer may be used in place of this diode and meter combination, but may not be so sensitive.) In this circuit, selectivity is improved by tapping the indicator circuit down coil L1 to minimize loading of the tuned circuit. But this calls for a 3-terminal plug-in coil. In Fig. 2-14C, the indicator is a small pilot lamp, such as the 2-V, 60-mA Type 48. This arrangement can be used only when the signal source—such as a radio transmitter or industrial oscillator—supplies enough power to light the lamp. In this arrangement, peak brilliance of the lamp indicates resonance.

Table 2-1 gives coil-winding data for a wavemeter employing a 140-pF tuning capacitor. The four inductors cover the frequency spectrum of 1.1 to 150 MHz in four bands. Other inductance and capacitance combinations may be worked out for other frequencies (see Equations 2-1, 2-2, and 2-3).

2.14 VARACTOR FREQUENCY MULTIPLIER

The varactor was introduced in Section 2.11 in its role as a dc-variable capacitor for resonant-circuit tuning. The most important large-signal application of the varactor in company with LC tuned circuits is harmonic generation. The latter property arises as a result of the pronounced distortion occurring when the varactor is operated over its entire range of nonlinear response. This property is utilized in modern high-efficiency passive frequency doublers, triplers, quadruplers, and higher-order multipliers in transmitters and other radio-frequency equipment.

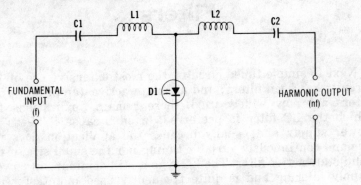

Fig. 2-15. Basic varactor multiplier.

Fig. 2-15 shows a basic varactor frequency multiplier circuit. Input (driving) current at the fundamental frequency (f) flows through the left loop (C1L1D1) of the circuit. Series-resonant circuit L1C1 is tuned to this frequency. Flow of this current through the varactor generates a number of harmonics. The desired harmonic (nf) is selected by the second series-resonant circuit (L2C2) and is delivered to the HARMONIC OUTPUT terminals.

Every varactor multiplier circuit is some adaptation of the simple arrangement shown in Fig. 2-15; some, for instance, employ parallel-resonant circuits. The varactor multiplier is not only efficient (P_o/P_i approaches 100% for doublers, where P_o is output power, and P_i is input power, both radio-frequency), but it also requires no local power supply. The only power required for its operation is supplied by the input signal itself.

CHAPTER 3

Filters

Next to simple tuned circuits, the most extensive use of LC circuits has been filters; and although active (amplifier-type) filters are now widely used, there remain applications for which the LC filter is preferred, often because it needs no power supply and usually because the application does not demand continuously variable tuning nor the small size of the solid-state active filter. The LC type is still dominant in power-supply filtering and is quite frequently used in interference filtering.

Filters are conveniently classified as *wave filters* (those which process signals) and *power-supply filters* (those which remove ripple from the dc output of a rectifier). This chapter describes each type and offers simple design data applicable to them.

3.1 BASIC FILTERING PROPERTIES OF L AND C

The inductor and the capacitor are both basically filters. Since their reactance changes with frequency, each transmits frequencies unequally, tending to separate some from others. This action is shown for the inductor in Fig. 3-1 and for the capacitor in Fig. 3-2.

These examples assume for illustrative purposes that the inductor and capacitor are pure reactance, i.e., that neither has resistance. Thus, in Fig. 3-1A a constant voltage of variable frequency is applied to inductor L and current meter M in series. The resulting current $I = E/X_L$. But X_L increases with

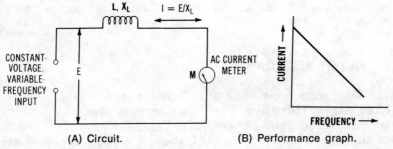

(A) Circuit.   (B) Performance graph.

Fig. 3-1. Basic filtering action of inductor.

frequency, so I decreases as the frequency is increased, and vice versa, as shown in Fig. 3-1B. (If $E = 1V$ and $L = 1$ H, current I is 1.59 mA at 100 Hz, 0.159 mA at 1000 Hz, and 159 μA at 10 kHz.) Similarly, in Fig. 3-2A the constant voltage of variable frequency is applied to capacitor C and current meter M in series. Here, the resulting current $I = E/X_c$. But X_c decreases with frequency, so I increases as the frequency is increased, and vice versa, as shown in Fig. 3-2B. (Here, by comparison, if $E = 1$ V and $C = 1$ μF, current I is 0.628 mA at 100 Hz, 6.28 mA at 1000 Hz, and 62.8 mA at 10 kHz.)

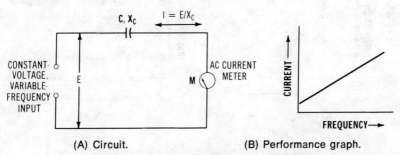

(A) Circuit.   (B) Performance graph.

Fig. 3-2. Basic filtering action of capacitor.

From Figs. 3-1B and 3-2B, respectively, it is seen that the inductor tends to *transmit* low frequencies and *attenuate* high frequencies, whereas the capacitor tends to attenuate low frequencies and transmit high frequencies. Now, while this simple filtering action is useful, it is limited, for there is no single point (*cutoff frequency*, f_c) on one side of which frequencies are transmitted with little loss, and on the other side of which frequencies are attenuated. The action is continuous in each case and uniform (Figs. 3-1B and 3-2B). When inductors and capacitors are combined, however, into *filter sections*, each enhances the filtering action of the other. Filter sections may be used singly or cascaded.

3.2 FILTER SECTIONS

A filter section is named, as to function: *low-pass, high-pass, bandpass,* or *bandstop* (the latter type is also called *band-suppression* or *band-elimination*). Fig. 3-3 shows *ideal* action of these sections. As to configuration, a section is named according to the Greek or Roman letter which its schematic resembles: *L, T,* or *pi.* Fig. 3-4 illustrates these basic configurations.

Figs. 3-5 to 3-15 give circuits and data for L-type low-pass, high-pass, bandpass, and bandstop filter sections. For each of these classes, two types are shown: *constant-k* and *m-derived* (both series and shunt type). The constant-k type is the simpler, but performance of the m type more closely ap-

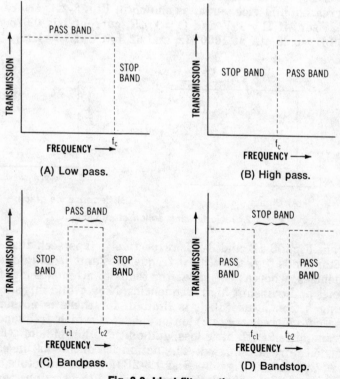

(A) Low pass.   (B) High pass.

(C) Bandpass.   (D) Bandstop.

Fig. 3-3. Ideal filter action.

proaches the ideal. The constant-k type is so called because the product of the impedance (Z1) of its series arm and the

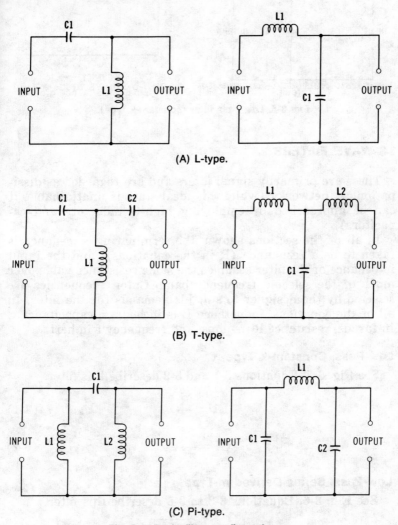

(A) L-type.

(B) T-type.

(C) Pi-type.

Fig. 3-4. Basic filter configurations.

impedance (Z2) of its shunt arm equals a constant: $Z1Z2 = k^2$. In the m type, the factor m governs the ratio of cutoff frequency f_c to a given high-attenuation frequency (the frequency, for instance, at which transmission approaches zero) and generally has the selected value 0.6.

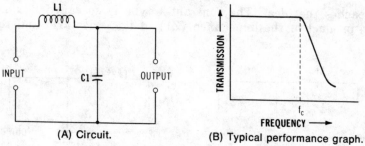

(A) Circuit.   (B) Typical performance graph.

Fig. 3-5. Low-pass filter (constant-k type).

3.3 WAVE FILTERS

These are primarily signal filters and are regarded as dissipationless networks (while this ideal state is unattainable, it can be approached by employing high-Q inductors and capacitors).

In all of the sections shown, the terminating impedance is taken to be a resistance (R in the equations), and the input impedance of the filter assumes the same resistance value over much of the selected frequency band. Cutoff frequencies are selected by the designer to suit his demands for the filter. In all of the equations, inductance L is in henrys, capacitance C in farads, resistance R in ohms, and frequency f in hertz.

Low-Pass, Constant-k Type

See Fig. 3-5. Equations 3-1 and 3-2 describe this filter.

$$L1 = \frac{R}{\pi f_c} \tag{3-1}$$

$$C1 = \frac{1}{\pi f_c R} \tag{3-2}$$

Low-Pass, Series-Derived m-Type

See Fig. 3-6. Equations 3-3 to 3-6 describe this filter.

$$m = \sqrt{1 - \frac{f_{c1}^2}{f_{c2}^2}} \tag{3-3}$$

$$L1 = \frac{mR}{\pi f_{c1}} \tag{3-4}$$

$$L2 = \frac{(1 - m^2)R}{4\pi m f_{c1}} \tag{3-5}$$

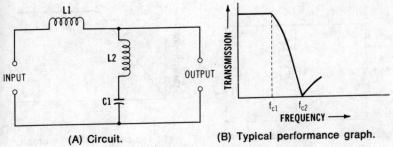

(A) Circuit.   (B) Typical performance graph.

Fig. 3-6. Low-pass filter (series-derived m-type).

$$C1 = \frac{1 - m^2}{\pi f_{c1}} \qquad (3\text{-}6)$$

Low-Pass, Shunt-Derived m-Type

See Fig. 3-7. Equations 3-7 to 3-10 describe this filter.

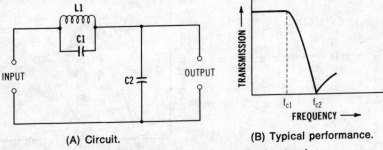

(A) Circuit.   (B) Typical performance.

Fig. 3-7. Low-pass filter (shunt-derived m-type).

$$m = \sqrt{1 - \frac{f_{c1}^2}{f_{c2}^2}} \qquad (3\text{-}7)$$

$$L1 = \frac{mR}{\pi f_{c1}} \qquad (3\text{-}8)$$

$$C1 = \frac{1 - m_2}{4\pi m f_{c1} R} \qquad (3\text{-}9)$$

$$C2 = \frac{m}{\pi f_{c2} R} \qquad (3\text{-}10)$$

High-Pass, Constant-k Type

See Fig. 3-8. Equations 3-11 and 3-12 describe this filter.

$$L1 = \frac{R}{4\pi f_c} \qquad (3\text{-}11)$$

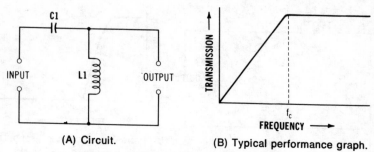

(A) Circuit.   (B) Typical performance graph.

Fig. 3-8. High-pass filter (constant-k type).

$$C1 = \frac{1}{4\pi f_c R} \qquad (3\text{-}12)$$

High-Pass, Series-Derived m-Type

See Fig. 3-9. Equations 3-13 to 3-16 describe this filter.

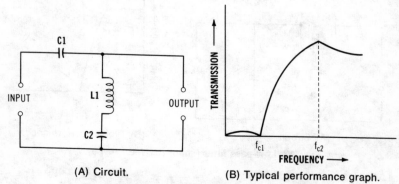

(A) Circuit.   (B) Typical performance graph.

Fig. 3-9. High-pass filter (series-derived m-type).

$$m = \sqrt{1 - \frac{f_{c1}{}^2}{f_{c2}{}^2}} \qquad (3\text{-}13)$$

$$L1 = \frac{R}{4\pi m f_{c2}} \qquad (3\text{-}14)$$

$$C1 = \frac{1}{4\pi m f_{c2} R} \qquad (3\text{-}15)$$

$$C2 = \frac{m}{(1 - m^2)\pi f_{c2} R} \qquad (3\text{-}16)$$

High-Pass, Shunt-Derived m-Type

See Fig. 3-10. Equations 3-17 to 3-20 describe this filter.

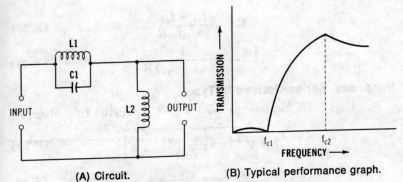

(A) Circuit.

(B) Typical performance graph.

Fig. 3-10. High-pass filter (shunt-derived m-type).

$$m = \sqrt{1 - \frac{f_{c1}^2}{f_{c2}^2}} \tag{3-17}$$

$$L1 = \frac{mR}{(1 - m^2)\pi f_{c2}} \tag{3-18}$$

$$L2 = \frac{R}{4\pi m f_{c2}} \tag{3-19}$$

$$C1 = \frac{1}{4\pi m f_{c2} R} \tag{3-20}$$

Bandpass, Constant-k Type

See Fig. 3-11. Equations 3-21 to 3-24 describe this filter.

$$L1 = \frac{R}{\pi(f_{c2} - f_{c1})} \tag{3-21}$$

$$L2 = \frac{f_{c2} - f_{c1}}{4\pi f_{c1} f_{c2}} \tag{3-22}$$

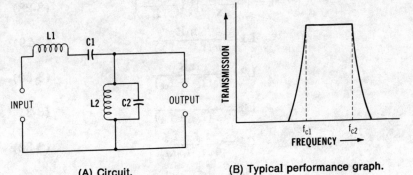

(A) Circuit.

(B) Typical performance graph.

Fig. 3-11. Bandpass filter (constant-k type).

$$C1 = \frac{f_{c2} - f_{c1}}{4\pi f_{c1}f_{c2}R} \tag{3-23}$$

$$C2 = \frac{1}{\pi(f_{c2} - f_{c1})R} \tag{3-24}$$

Bandpass, Series-Derived m-Type

See Fig. 3-12. Equations 3-25 to 3-34 describe this filter.

$$x = \sqrt{\left(1 - \frac{f_{c2}^2}{f_{c3}^2}\right)\left(1 - \frac{f_{c3}^2}{f_{c4}^2}\right)} \tag{3-25}$$

$$m = \frac{x}{1 - \left(\frac{f_{c2}f_{c3}}{f_{c4}^2}\right)} \tag{3-26}$$

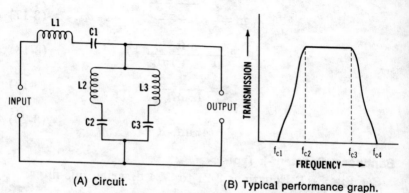

(A) Circuit.   (B) Typical performance graph.

Fig. 3-12. Bandpass filter (series-derived m-type).

$$y = \frac{(1 - m^2)f_{c2}f_{c3}}{4f_{c1}^2x}\left(1 - \frac{f_{c1}^2}{f_{c4}^2}\right) \tag{3-27}$$

$$z = \left(\frac{1 - m^2}{4x}\right)\left(1 - \frac{f_{c1}^2}{f_{c4}^2}\right) \tag{3-28}$$

$$L1 = \frac{mR}{\pi(f_{c3} - f_{c2})} \tag{3-29}$$

$$L2 = \frac{zR}{\pi(f_{c3} - f_{c2})} \tag{3-30}$$

$$L3 = \frac{yR}{\pi(f_{c3} - f_{c2})} \tag{3-31}$$

$$C1 = \frac{f_{c3} - f_{c2}}{4\pi f_{c2}f_{c3}mR} \tag{3-32}$$

$$C2 = \frac{f_{c3} - f_{c2}}{4\pi f_{c2}f_{c3}yR} \tag{3-33}$$

$$C3 = \frac{f_{c3} - f_{c2}}{4\pi f_{c2}f_{c3}zR} \tag{3-34}$$

Bandpass, Shunt-Derived m-Type

See Fig. 3-13. Equations 3-35 to 3-40 describe this filter. (For x, y, z, and m, see Equations 3-25, 3-27, and 3-28. To identify all f_c's, see Fig. 3-13B.)

$$L1 = \frac{(f_{c3} - f_{c2})R}{4\pi f_{c2}f_{c3}z} \tag{3-35}$$

$$L2 = \frac{(f_{c3} - f_{c2})R}{4\pi f_{c2}f_{c3}y} \tag{3-36}$$

$$L3 = \frac{(f_{c3} - f_{c2})R}{4\pi f_{c2}f_{c3}m} \tag{3-37}$$

$$C1 = \frac{y}{\pi(f_{c3} - f_{c2})R} \tag{3-38}$$

$$C2 = \frac{y}{\pi(f_{c3} - f_{c2})R} \tag{3-39}$$

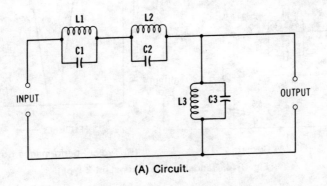

(A) Circuit.

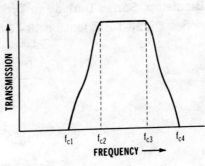

(B) Typical performance graph.

Fig. 3-13. Bandpass filter (shunt-derived m-type).

$$C3 = \frac{m}{\pi (f_{c3} - f_{c2}) R} \tag{3-40}$$

Bandstop, Constant-k Type

See Fig. 3-14. Equations 3-41 to 3-45 describe this filter. (In Equation 3-41 and Fig. 3-14B, f_m is the center frequency of the stopband.)

$$f_m = \sqrt{f_{c1} f_{c2}} \tag{3-41}$$

$$L1 = \frac{(f_{c2} - f_{c1}) R}{\pi f_{c1} f_{c2}} \tag{3-42}$$

$$L2 = \frac{R}{4\pi (f_{c2} - f_{c1})} \tag{3-43}$$

$$C1 = \frac{1}{4\pi (f_{c2} - f_{c1}) R} \tag{3-44}$$

$$C2 = \frac{f_{c2} - f_{c1}}{4\pi (f_{c1} f_{c2}) R} \tag{3-45}$$

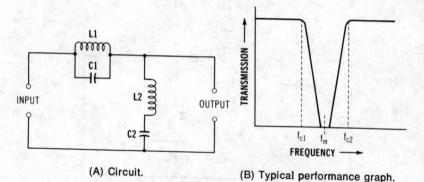

(A) Circuit.   (B) Typical performance graph.

Fig. 3-14. Bandstop filter (constant-k type).

Bandstop, Series-Derived m-Type

See Fig. 3-15. Equations 3-46 to 3-54, describe this filter.

$$m = \sqrt{\frac{\left(1 - \frac{f_{c1}^2}{f_{c3}^2}\right)\left(1 - \frac{f_{c3}^2}{f_{c4}^2}\right)}{1 - \frac{f_{c1}}{f_{c4}}}} \tag{3-46}$$

$$x = \frac{1}{m}\left(1 + \frac{f_{c1} f_{c4}}{f_{c3}^2}\right) \tag{3-47}$$

$$y = \frac{1}{m}\left(1 + \frac{f_{c3}^2}{f_{c1} f_{c4}}\right) \tag{3-48}$$

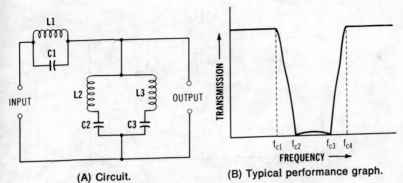

(A) Circuit.   (B) Typical performance graph.

Fig. 3-15. Bandstop filter (series-derived m-type).

$$L1 = \frac{m(f_{c4} - f_{c1})R}{\pi f_{c1} f_{c4}} \qquad (3\text{-}49)$$

$$L2 = \frac{R}{4\pi(f_{c4} - f_{c1})m} \qquad (3\text{-}50)$$

$$L3 = \frac{yR}{4\pi(f_{c4} - f_{c1})} \qquad (3\text{-}51)$$

$$C1 = \frac{1}{4\pi(f_{c4} - f_{c1})mR} \qquad (3\text{-}52)$$

$$C2 = \frac{f_{c4} - f_{c1}}{\pi f_{c1} f_{c4} yR} \qquad (3\text{-}53)$$

$$C3 = \frac{f_{c4} - f_{c1}}{\pi f_{c1} f_{c4} Rx} \qquad (3\text{-}54)$$

Bandstop, Shunt-Derived m-Type

See Fig. 3-16. Equations 3-55 to 3-60 describe this filter. (For m, x, and y, see Equations 3-46, 3-47, and 3-48. To identify all f_c's, see Fig. 3-16B.)

$$L1 = \frac{(f_{c4} - f_{c1})R}{\pi f_{c1} f_{c4} y} \qquad (3\text{-}55)$$

$$L2 = \frac{(f_{c4} - f_{c1})R}{\pi f_{c1} f_{c4} x} \qquad (3\text{-}56)$$

$$L3 = \frac{R}{4\pi(f_{c4} - f_{c1})m} \qquad (3\text{-}57)$$

$$C1 = \frac{x}{4\pi(f_{c4} - f_{c1})R} \qquad (3\text{-}58)$$

$$C2 = \frac{y}{4\pi(f_{c4} - f_{c1})R} \qquad (3\text{-}59)$$

71

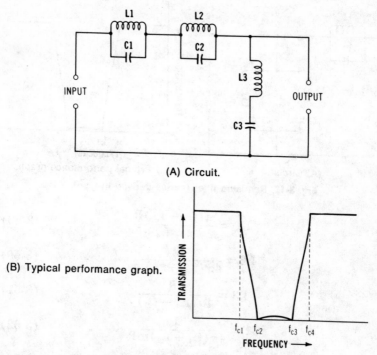

(A) Circuit.

(B) Typical performance graph.

Fig. 3-16. Bandstop filter (shunt-derived m-type).

$$C3 = \frac{m(f_{c4} - f_{c1})}{\pi f_{c1} f_{c4} R} \qquad (3\text{-}60)$$

3.4 POWER-SUPPLY FILTERS

The purpose of a power-supply filter is to remove the ripple from the dc output of a rectifier. In such a filter, the inductor is the series arm and the capacitor is the shunt arm. The choke blocks the flow of the ripple, hence, its familiar name of *choke coil;* the capacitor short-circuits the ripple (which is an ac component) to ground, hence, its familiar name of *bypass*. In this action, the choke cannot block the dc component, and the capacitor cannot short-circuit the dc component.

The power-supply filter is a low-pass circuit, generally of the pi or L configuration. For best results, it should be designed for a cutoff frequency considerably lower than the ripple frequency (see Table 3-1), to ensure that the ripple and its harmonics all will be severely attenuated. The choke must be capable of handling the maximum direct-current load

Table 3-1. Ripple Frequency for Common Power Supplies

Type of Supply	Ripple Frequency (f = line-voltage frequency)
Single-Phase, Half-Wave	f
Single-Phase, Center-Tap	2f
Single-Phase Bridge	2f
3-Phase Star	3f
3-Phase Delta	6f
3-Phase, Full-Wave, Single-Y	6f

of the supply, and the capacitor(s) must be able to operate safely at the peak output voltage of the rectifier.

Fig. 3-17 illustrates the two basic types of single-section power-supply filter. The capacitor-input type (Fig. 3-17A) provides the higher dc output voltage, which at low output-current level (high load resistance) may approximate the peak value of the ac voltage input to the rectifier (i.e., $E_{dc} = 1.41 \, E_{ac \, rms}$), but has the poorer voltage regulation. The choke-input type (Fig. 3-17B) provides lower dc voltage (E_{dc} = approximately $0.9 E_{ac \, rms}$), but has superior output-voltage regulation. Note that in each instance an output capacitor is required (C2 in Fig. 3-17A, C1 in Fig. 3-17B). Fig. 3-18 shows typical

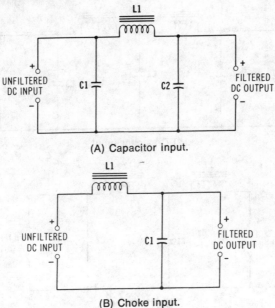

(A) Capacitor input.

(B) Choke input.

Fig. 3-17. Typical single-section power-supply filters.

two-section filters for increased smoothing of dc output. In power-supply filters, capacitors function not only as frequency-selective elements, but also as energy-storage devices: during the rise of the rectified-voltage pulse from zero to maximum, the capacitor charges while current is being delivered to the dc output load; then, as the voltage subsequently decreases to zero, the capacitor discharges, delivering its energy to the output load and thus maintaining the filter output voltage constant.

A cursory inspection of Figs. 3-17 and 3-18 shows that the filters all are low-pass, constant-k type. Their design accordingly may be accomplished with the aid of Equations 3-1 and 3-2. (Load R in these formulas will be an actual resistance to which the power supply delivers energy or will be the quotient E/I, where E is the dc load voltage in volts, and I is the dc load current in amperes.) For some applications, the design procedure is bypassed completely. Instead, on-hand high-inductance chokes and high-capacitance capacitors simply are teamed up. The resulting cutoff frequency generally is so low that the filter is effective in a large number of possible applications. Such an arrangement is familiarly termed a *brute-force filter*.

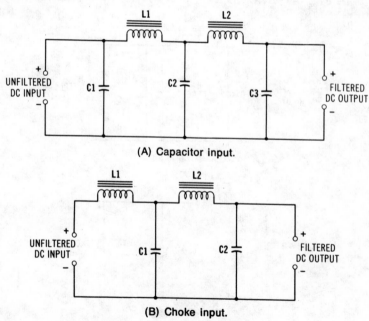

(A) Capacitor input.

(B) Choke input.

Fig. 3-18. Typical cascaded filters.

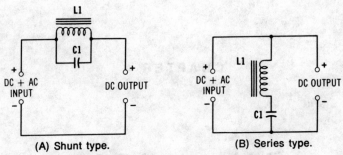

(A) Shunt type.　　　　　　(B) Series type.

Fig. 3-19. Resonant single-frequency filters.

Occasionally, a single frequency—such as a troublesome harmonic of the ac-supply frequency—must be removed from the output of a rectifier. Simple LC combinations for performing this task are shown in Fig. 3-19. In Fig. 3-19A, a shunt L1C1 circuit traps the signal; in Fig. 3-19B, a series L1C1 circuit short-circuits the signal to ground. These arrangements are seldom satisfactory as the complete filter for a power supply, since they remove only one frequency (at best a very narrow band of frequencies), but they are very effective when used in conjunction with a regular power-supply filter. Both are immediately recognized as the simple wavetraps described in Section 2.12, Chapter 2.

CHAPTER 4

Bridges and Other
Measuring Devices

Inductance-capacitance combinations are often seen in oscillators and signal generators, where they function as tanks and filters, and in special-purpose meters, where they serve as selective filters and as tuners. They are also the basis of other test equipment in which an unknown inductance is compared with a known capacitance, such as ac bridges (in combination with resistance), frequency meters (e.g., the absorption wavemeter already described in Section 2.13, Chapter 2), and component testers using the resonance method.

This chapter describes several of the better-known LC-based test and measuring devices.

4.1 ANDERSON BRIDGE

See Fig. 4-1. This is a six-impedance network (L_x, C_s, R1, R2, R3, R4) in which unknown inductance L_x is compared with standard capacitance C_s. Although harder to adjust than the more common four-impedance bridges, this circuit offers a wider range.

The inductance balance (R3) and Q balance (R1) are independent of each other and of the generator frequency. At null:

$$L_x = C_s[R3(1 + R2/R4) + R2] \qquad (4\text{-}1)$$

where,

L_x is in henrys,
C_s is in farads,
All Rs are in ohms.

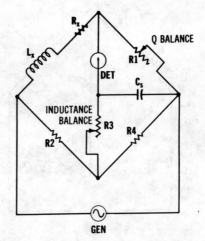

Fig. 4-1. Anderson bridge.

The equivalent resistance (R_x) of the inductor under test is calculated:

$$R_x = \frac{(R1R2)}{R4} \qquad (4\text{-}2)$$

4.2 HAY BRIDGE

See Fig. 4-2. In this bridge, unknown inductance L_x is compared with standard capacitance C_s. The circuit is balanced in the conventional manner by alternate adjustment of inductance-balance rheostat R2 and Q-balance rheostat R3. At null:

$$L_x = \frac{R1R2R3}{1 + (R3^2\omega 2C_s^2)} \qquad (4\text{-}3)$$

where,

    L_x is in henrys,
    C_s is in farads,
    ω is $2\pi f$ (f is in hertz),
    All Rs are in ohms.

Since R_3 appears in the denominator of the fraction, the inductance balance is not independent of the Q balance. Also, since ω also appears in the equation, the balance is frequency dependent. However, if the Q of the inductor under test is higher than 10, the frequency may be ignored, and Equation 4-3 may be simplified with an error of less than 1 percent:

$$L_x = C_s(R1R2) \qquad (4\text{-}4)$$

77

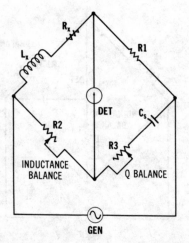

Fig. 4-2. Hay bridge.

At null, the equivalent resistance (R_x) of the inductor under test is calculated:

$$R_x = \frac{\omega^2 C_s^2 R1R2R3}{1 + (\omega^2 C_s^2 R3^2)} \tag{4-5}$$

Again note that the settings of both the inductance-balance rheostat (R2) and Q-balance rheostat (R3) enter into the calculation and that the R_x determination, like that of L_x, is frequency dependent. However, if the Q of the inductor is higher than 10, Equation 4-5 may be simplified:

$$R_x = \frac{R1R2}{R3} \tag{4-6}$$

4.3 MAXWELL BRIDGE

See Fig. 4-3. This circuit compares unknown inductance L_x with standard capacitor C_s. It differs from the Hay bridge described in the preceding section in the parallel connection of the Q-balance rheostat (R3) and standard capacitor (C_s).

In this bridge, the inductance balance and Q balance are independent of each other and each is independent of frequency. At null:

$$L_x = C_s (R1R2) \tag{4-7}$$

where,

L_x is in henrys,
C_s is in farads,
Both Rs are in ohms.

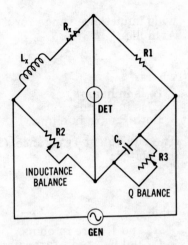

Fig. 4-3. Maxwell bridge.

The equivalent resistance (R_x) of the inductor under test is calculated:

$$R_x = R1 \left(\frac{R2}{R3}\right) \qquad (4\text{-}8)$$

4.4 OWEN BRIDGE

See Fig. 4-4. In this circuit, a variable capacitor (C_r) is employed as the Q-balance component. The inductance balance and Q balance are independent of each other and each is independent of frequency. The Owen bridge provides an extremely

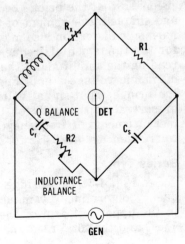

Fig. 4-4. Owen bridge.

wide inductance range for a narrow range of C_s and R values. At null:

$$L_x = C_s(R1R2) \tag{4-9}$$

where,
   L_x is in henrys,
   C_s is in farads,
   Both Rs are in ohms.

The equivalent resistance (R_x) of the inductor under test is calculated:

$$R_x = R1 \left(\frac{C_s}{C_r}\right) \tag{4-10}$$

where,
   R_x and R1 are in ohms,
   C_r and C_s are in farads.

4.5 RESONANCE BRIDGES

While frequency sensitivity can be somewhat of a nuisance in ac bridges used for inductance, capacitance, and resistance measurements, this property may be utilized for the measurement of frequency if all of the bridge arms are filled and the bridge then is balanced for frequency. In such an instance, the balance-adjustment component (rheostat or variable capacitor) may be made direct reading in frequency instead of in R, C, or L. The resistance, capacitance, and/or inductance values in the bridge arms usually can be known with high precision, so that the bridge becomes a useful device for accurate and simple measurement of frequency. Measurements of total harmonic distortion also may be made by balancing the bridge for the fundamental frequency of a complex test signal, then reading the amplitude of the output at null (which is the total harmonic voltage), and, finally, comparing this voltage with the bridge input-signal voltage. The resonance bridge is one of several types used for these purposes. Resonance-bridge measurements of frequency and distortion are usually limited to frequencies of 20 Hz to 20 kHz.

Series Type

See Fig. 4-5. This bridge may be balanced only at the frequency determined by inductance L1 and the capacitance C1 in the upper left arm of the circuit. At this frequency, $\omega L = 1/\omega C$ and the total reactance = 0, and only the internal resis-

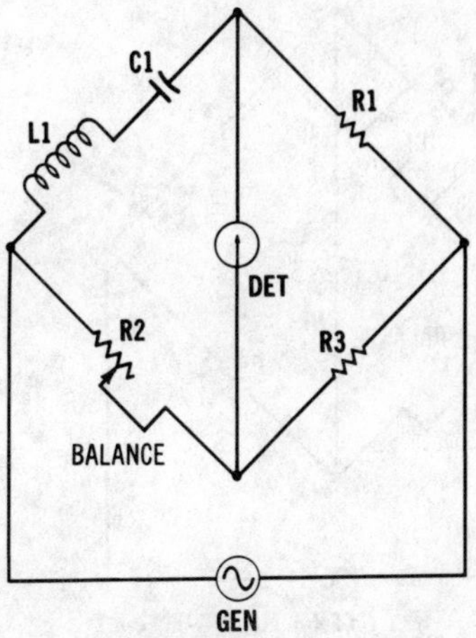

Fig. 4-5. Resonant bridge
(series type).

tance of the inductor and capacitor remain in that arm, making the circuit a four-arm resistance bridge at that one frequency. At null, the unknown frequency is:

$$f_x = \frac{1}{2\pi\sqrt{L1C1}} \qquad (4\text{-}11)$$

where,
  f_x is in hertz,
  L1 is in henrys,
  C1 is in farads,
  π is 3.1416.

A desired frequency spectrum may be covered by making C1 variable for tuning, and by switching L1 to appropriate values to change ranges. (Since high capacitances are needed for the audio spectrum where the resonance bridge is effective, a variable C1 would for practical purposes be a capacitor decade.) For sharp null response, both C1 and L1 must be high-Q components.

Shunt-Type

See Fig. 4-6. Because the impedance arm in the series-type circuit (Fig. 4-5) is a series-resonant circuit, the current through this arm and the one containing resistor R1 is maximum at the resonant frequency and can reach high levels. With some inductors, especially when high input-signal voltage and low values of R1 are unavoidable, the resulting high current can introduce enough distortion to obscure the null. To eliminate these difficulties, the impedance arm has been changed in Fig. 4-6 to a shunt connection of the inductance and

81

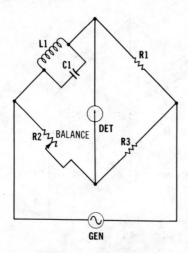

Fig. 4-6. Resonance bridge
(shunt type).

capacitance (parallel-resonant circuit). In this arrangement, the current through the impedance arm and R1 is minimum at resonance, approaching zero for high-Q inductors and capacitors. In all other respects, the shunt-type bridge is identical to the series type described in the preceding section. The balance equation likewise is the same (see Equation 4-11).

4.6 BRIDGED-T NULL NETWORK

The circuit shown in Fig. 4-7A provides a sharp null when high-Q inductor and capacitors are employed. It takes its name from the fact that inductor L1 bridges the T network formed by capacitors C1 and C2 and resistor R1. In the inductive arm of the circuit, R_x is the equivalent series resistance of the inductor. Capacitances C1 and C2 are made equal and may be varied simultaneously to tune the network. (Alternatively, the capacitors may be held at a fixed value, and the inductance varied.) To balance the network, the capacitances (or inductance) and resistance R1 must be adjusted alternately, C1 being equal to C2 at all points. At null:

$$f_n = \frac{1}{\pi\sqrt{2L1C1}}$$ (4-12)

where,
 f_n is in hertz,
 L1 is in henrys,
 C1 is in farads,
 π is 3.1416.

82

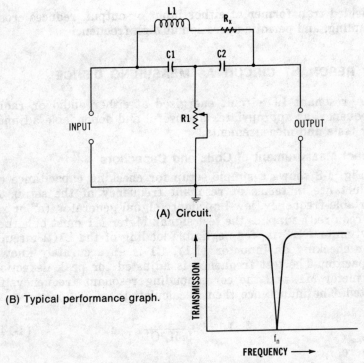

(A) Circuit.

(B) Typical performance graph.

Fig. 4-7. Bridge null network.

The equivalent resistance (R_x) of inductor L1 is calculated:

$$R_x = \frac{1}{R1(\omega C1)^2} \qquad (4\text{-}13)$$

where,
R_x and R1 are in ohms,
C1 is in farads,
ω is $2\pi f$ (f is in hertz).

The bridged-T network is used for frequency measurements and distortion measurements in the same manner as explained for the resonance bridge (Section 4.5). It may be used also for capacitance measurement (C1 or C2 = $1/8\pi^2 f^2 L$), or inductance measurement (L1 = $1/2\pi^2 f^2 C1$). In this function, it is used as high as the lower vhf spectrum. This device also finds wide use as a bandstop filter (*notch filter*). An advantage of the bridged-T network over an equivalent bridge, when it can be used, is its provision of a common ground for generator, network, and detector. This removes the need for a shielded transformer at either input or output, reduces cross coupling, and permits operation at high frequencies.

4.7 RESONANT CIRCUIT AS MEASURING DEVICE

A resonant LC circuit energized at either audio or radio frequency, as appropriate, is a useful and dependable adjunct for tests and measurements.

Direct Measurement of Coils and Capacitors

Fig. 4-8 shows a simple setup for checking capacitance or inductance in terms of resonant frequency of the setup. A variable-frequency, low-impedance signal generator (af or rf, as required) supplies the test signal. Meter M1 must be either a vtvm or FETvm, for negligible loading of the L1C1 circuit. For checking an inductor (L1), C1 is an accurately known capacitor. The test frequency is adjusted for peak deflection of meter M1, and the corresponding resonant frequency, f_r, noted. The inductance then is calculated:

$$L1 = \frac{1}{39.5f_r^2C1} \tag{4-14}$$

where,
   L1 is in henrys,
   f_r is in hertz,
   C1 is in farads.

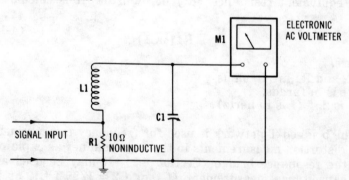

Fig. 4-8. Resonant circuit for component testing.

For checking a capacitor (C1), L1 is an accurately known inductor. The test frequency is adjusted for peak deflection of meter M1, and the corresponding resonant frequency, f_r, noted. The capacitance then is calculated:

$$C1 = \frac{1}{39.5f_r{}^2L1} \qquad (4\text{-}15)$$

where,
  C1 is in farads,
  f_r is in hertz,
  L1 is in henrys.

This direct method of measurement does not take into account the distributed capacitance of the inductor, nor stray circuit capacitance. When high accuracy is desired, it is necessary to add these components to C1, and usually they are difficult to find. The substitution method automatically compensates for distributed and wiring capacitances.

Substitution Method for Capacitors

See Fig. 4-9. This setup is similar to Fig. 4-8, except that here the capacitor under test is connected to terminals X1 and X2 to place it in parallel with variable capacitor C1. The latter has a dial reading directly in picofarads. *Test Procedure:* (1) With terminals X1 and X2 open, set tuning capacitor C1 to its maximum-capacitance position: C1$_a$. (2) Adjust the test frequency for peak deflection of meter M1. (3) Using the shortest practicable leads, connect the test capacitor (C_x) to terminals X1 and X2. This additional capacitance

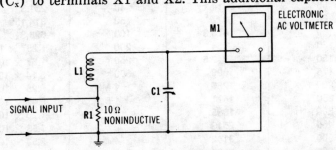

Fig. 4-9. Resonant circuit for substitution-type capacitance measurement.

detunes the circuit, causing the meter reading to fall. (4) Detune C1 to restore peak deflection. (5) At this point, read the new capacitance setting: C1$_b$. (6) Calculate the unknown capacitance:

$$C_x = C1_a - C1_b \qquad (4\text{-}16)$$

The substitution method has the disadvantage that it is limited to capacitances that do not exceed the maximum capacitance (C1$_a$) of tuning capacitor C1.

APPENDIX A

Angular Velocity

f	ω	f	ω
1 Hz	6.28	5000	31,416
10	62.8	10 kHz	62,832
20	125.7	12	75,398
25	157.1	15	94,248
30	188.5	20	125,664
40	251.3	25	157,081
50	314.1	40	251,327
60	377	50	314,159
100	628.3	100	628,318
120	754	200	1.257×10^6
150	942.5	250	1.57×10^6
175	1099	500	3.14×10^6
200	1256	1 MHz	6.28×10^6
250	1571	1.5	9.42×10^6
300	1885	2	1.26×10^7
400	2513	3	1.88×10^7
500	3142	4	2.51×10^7
1000	6283	5	3.14×10^7
1250	7854	10	6.28×10^7
1500	9425	15	9.42×10^7
1600	10,053	20	1.26×10^8
2000	12,566	25	1.57×10^8
2500	15,708	50	3.14×10^8
3000	18,849	100	6.28×10^8
4000	25,133		

Reactance of Inductors at 1000 Hz

Inductance (L)	Reactance (X_L in ohms)
1 μH	0.00628
10 μH	0.0628
100 μH	0.628
1000 μH (1 mH)	6.28
10 mH	62.8
100 mH	628
1000 mH (1 H)	6283
10 H	62,831
100 H	628,318
1000 H	6.28×10^6

(1) For a frequency (f_x) other than 1000 Hz, multiply the 1000-Hz reactance by $f_x/1000$.
(2) Interpolate for intermediate values of inductance.

APPENDIX C

Reactance of Capacitors

For a frequency (f in hertz) other than 1000 Hz, multiply the 1000-Hz reactance by $1/f$.

Capacitance (C)	Reactance (X_c) (ohms)	Capacitance (C)	Reactance (X_c) (ohms)
5 pF	3.18×10^7	150 pF	1.061×10^6
6.8 pF	2.338×10^7	180 pF	884,194
10 pF	1.591×10^7	200 pF	795,775
12 pF	1.326×10^7	220 pF	723,431
15 pF	1.061×10^7	250 pF	636,620
18 pF	8.842×10^6	270 pF	589,463
20 pF	7.958×10^6	300 pF	530,516
22 pF	7.234×10^6	330 pF	482,287
25 pF	6.366×10^6	390 pF	408,089
27 pF	5.895×10^6	470 pF	338,627
30 pF	5.305×10^6	500 pF	318,309
33 pF	4.823×10^6	560 pF	284,205
39 pF	4.081×10^6	0.001 μF	159,155
47 pF	3.386×10^6	0.0012 μF	133,629
50 pF	3.183×10^6	0.0015 μF	106,103
56 pF	2.842×10^6	0.0018 μF	88,419
68 pF	2.340×10^6	0.002 μF	79,578
75 pF	2.122×10^6	0.0022 μF	72,343
82 pF	1.941×10^6	0.0025 μF	63,662
100 pF	1.591×10^6	0.0027 μF	58,946
120 pF	1.326×10^6	0.003 μF	53,052

Capacitance (C)	Reactance (X_c) (ohms)	Capacitance (C)	Reactance (X_c) (ohms)
0.0033 μF	48,228	2.3 μF	69.2
0.0039 μF	40,809	3.3 μF	48.2
0.004 μF	39,788	4 μF	39.8
0.0047 μF	33,863	4.18 μF	38.1
0.005 μF	31,831	4.67 μF	34.1
0.0056 μF	28,420	4.7 μF	33.9
0.006 μF	26,526	4.88 μF	32.6
0.0068 μF	23,405	5 μF	31.8
0.007 μF	22,736	5.1 μF	31.2
0.0082 μF	19,409	8 μF	19.9
0.01 μF	15,591	10 μF	15.9
0.012 μF	13,263	15 μF	10.6
0.015 μF	10,610	16 μF	9.95
0.02 μF	7958	20 μF	7.96
0.022 μF	7234	22 μF	7.23
0.025 μF	6366	25 μF	6.37
0.027 μF	5895	30 μF	5.30
0.03 μF	5305	33 μF	4.82
0.033 μF	4823	40 μF	3.98
0.035 μF	4547	47 μF	3.39
0.039 μF	4081	50 μF	3.18
0.04 μF	3979	60 μF	2.65
0.047 μF	3386	70 μF	2.27
0.05 μF	3183	75 μF	2.12
0.056 μF	2842	80 μF	1.99
0.058 μF	2744	100 μF	1.59
0.06 μF	2652	150 μF	1.06
0.068 μF	2340	160 μF	0.995
0.082 μF	1941	200 μF	0.796
0.086 μF	1851	220 μF	0.723
0.1 μF	1591	250 μF	0.637
0.15 μF	1061	300 μF	0.530
0.2 μF	796	330 μF	0.482
0.22 μF	723	470 μF	0.339
0.25 μF	637	500 μF	0.318
0.33 μF	482	1000 μF	0.159
0.47 μF	339	1500 μF	0.106
0.5 μF	318	2000 μF	0.0796
0.68 μF	234	2200 μF	0.0723
1 μF	159	3300 μF	0.0482
2 μF	79.6	4000 μF	0.0398
2.2 μF	72.3	5000 μF	0.0318

RC Time Constants

Capacitance (μF)	Resistance			
	1Ω	10Ω	100Ω	1000Ω
0.0001	10^{-10}	10^{-9}	10^{-8}	10^{-7}
0.001	10^{-9}	10^{-8}	10^{-7}	10^{-6}
0.01	10^{-8}	10^{-7}	10^{-6}	10^{-5}
0.1	10^{-7}	10^{-6}	10^{-5}	10^{-4}
1	10^{-6}	10^{-5}	10^{-4}	0.001
10	10^{-5}	10^{-4}	0.001	0.01
100	10^{-4}	0.001	0.01	0.1
1000	0.001	0.01	0.1	1

	Resistance			
	10k	100k	1MΩ	10MΩ
0.0001	10^{-6}	10^{-5}	10^{-4}	0.001
0.001	10^{-5}	10^{-4}	0.001	0.01
0.01	10^{-4}	0.001	0.01	0.1
0.1	0.001	0.01	0.1	1
1	0.01	0.1	1	10
10	0.1	1	10	100
100	1	10	100	1000
1000	10	100	1000	10,000

NOTE: Time constants given above are in seconds.
Interpolate for intermediate capacitance and/or resistance values.

APPENDIX E

RL Time Constants

Inductance (henrys)	Resistance (ohms)							
	1	10	100	1K	10K	100K	1M	10M
0.0001	0.0001	0.00001	1×10^{-6}	1×10^{-7}	1×10^{-8}	1×10^{-9}	1×10^{-10}	1×10^{-11}
0.001	0.001	0.0001	0.00001	1×10^{-6}	1×10^{-7}	1×10^{-8}	1×10^{-9}	1×10^{-10}
0.01	0.01	0.001	0.0001	0.00001	1×10^{-6}	1×10^{-7}	1×10^{-8}	1×10^{-9}
0.1	0.1	0.01	0.001	0.0001	0.00001	1×10^{-6}	1×10^{-7}	1×10^{-8}
1.0	1.0	0.1	0.01	0.001	0.0001	0.00001	1×10^{-6}	1×10^{-7}
10	10	1.0	0.1	0.01	0.001	0.0001	0.00001	1×10^{-6}
100	100	10	1.0	0.1	0.01	0.001	0.0001	0.00001
1000	1000	100	10	1.0	0.1	0.01	0.001	0.0001

NOTE: Time constants given above are in seconds. Interpolate for intermediate inductance and/or resistance values.

Resonant Frequency of LC Combinations

Resonant Frequency

Inductance → Capacitance ↓	10 pF	100 pF	1000 pF (0.001 μF)	0.01 μF	0.1 μF	1 μF	10 μF	100 μF	1000 μF
1 μH	50.3 MHz	15.9 MHz	5.03 MHz	1.59 MHz	503 kHz	159 kHz	50.3 kHz	15.9 kHz	5.03 kHz
10 μH	15.9 MHz	5.03 MHz	1.59 MHz	503 kHz	159 kHz	50.3 kHz	15.9 kHz	5.03 kHz	1591 Hz
100 μH	5.03 MHz	1.59 MHz	503 kHz	159 kHz	50.3 kHz	15.9 kHz	5.03 kHz	1591 Hz	503 Hz
1000 μH (1 mH)	1.59 MHz	503 kHz	159 kHz	50.3 kHz	15.9 kHz	5.03 kHz	1591 Hz	503 Hz	159 Hz
10 mH	503 kHz	159 kHz	50.3 kHz	15.9 kHz	5.03 kHz	1591 Hz	503 Hz	159 Hz	50.3 Hz
100 mH	159 kHz	50.3 kHz	15.9 kHz	5.03 kHz	1591 Hz	503 Hz	159 Hz	50.3 Hz	15.9 Hz
1000 mH (1H)	50.3 kHz	15.9 kHz	5.03 kHz	1591 Hz	503 Hz	159 Hz	50.3 Hz	15.9 Hz	5.03 Hz
10 H	15.9 kHz	5.03 kHz	1591 Hz	503 Hz	159 Hz	50.3 Hz	15.9 Hz	5.03 Hz	1.59 Hz
100 H	5.03 kHz	1591 Hz	503 Hz	159 Hz	50.3 Hz	15.9 Hz	5.03 Hz	1.59 Hz	0.503 Hz
1000 H	1591 Hz	503 Hz	159 Hz	50.3 Hz	15.9 Hz	5.03 Hz	1.59 Hz	0.503 Hz	0.159 Hz

APPENDIX G

Conversion Factors

To Convert From	To	Multiply by
Degrees	Grads	1.111
Degrees	Minutes	60
Degrees	Radians	0.0174533
Degrees	Seconds	3600
Farads	Microfarads	1×10^6
Farads	Nanofarads	1×10^9
Farads	Picofarads	1×10^{12}
Henrys	Microhenrys	1×10^6
Henrys	Millihenrys	1000
Henrys	Nanohenrys	1×10^9
Henrys	Picohenrys	1×10^{12}
Hertz	Kilohertz	0.001
Hertz	Megahertz	1×10^{-6}
Kilohertz	Hertz	1000
Kilohertz	Megahertz	0.001
Kilohms	Megohms	0.001
Kilohms	Ohms	1000
Megahertz	Hertz	1×10^6
Megahertz	Kilohertz	1000
Megohms	Kilohms	1000
Megohms	Ohms	1×10^6
Microfarads	Farads	1×10^{-6}
Microfarads	Nanofarads	1000
Microfarads	Picofarads	1×10^6
Microhenrys	Henrys	1×10^{-6}
Microhenrys	Millihenrys	0.001
Microhenrys	Nanohenrys	1000
Microhenrys	Picohenrys	1×10^6
Microseconds	Milliseconds	0.001

To Convert From	To	Multiply by
Microseconds	Minutes	1.667×10^{-8}
Microseconds	Seconds	1×10^{-6}
Millihenrys	Henrys	0.001
Millihenrys	Microhenrys	1×10^{3}
Millihenrys	Nanohenrys	1×10^{6}
Millihenrys	Picohenrys	1×10^{9}
Milliseconds	Microseconds	1000
Milliseconds	Minutes	1.667×10^{-5}
Milliseconds	Seconds	0.001
Minutes (angle)	Degrees	0.01667
Minutes (angle)	Grads	0.01852
Minutes (angle)	Radians	0.000291
Minutes (angle)	Seconds	60
Minutes (time)	Hours	0.01667
Minutes (time)	Microseconds	6×10^{7}
Minutes (time)	Milliseconds	60,000
Minutes (time)	Seconds	60
Nanofarads	Farads	1×10^{-9}
Nanofarads	Microfarads	0.001
Nanofarads	Picofarads	1000
Nanohenrys	Henrys	1×10^{-9}
Nanohenrys	Microhenrys	0.001
Nanohenrys	Millihenrys	1×10^{-6}
Nanohenrys	Picohenrys	1000
Ohms	Kilohms	0.001
Ohms	Megohms	1×10^{-6}
Picofarads	Farads	1×10^{-12}
Picofarads	Microfarads	1×10^{-6}
Picohenrys	Henrys	1×10^{-12}
Picohenrys	Microhenrys	1×10^{-6}
Picohenrys	Millihenrys	1×10^{-9}
Picohenrys	Nanohenrys	0.001
Radians	Degrees	57.2958
Radians	Grads	63.654
Radians	Minutes	3437.75
Radians	Seconds	206,265
Seconds (angle)	Degrees	0.000278
Seconds (angle)	Grads	0.0003086
Seconds (angle)	Minutes	0.01667
Seconds (angle)	Radians	4.848×10^{-6}
Seconds (time)	Hours	0.0002777
Seconds (time)	Microseconds	1×10^{6}
Seconds (time)	Milliseconds	1000
Seconds (time)	Minutes	0.016667